This volume contains the Proceedings of a Conference on Complex Analysis and Algebraic Geometry held in Göttingen during the summer of 1985. Its main purpose was to combine the methods of real analysis and algebraic geometry in the research of the field of Several Complex Variables. The Conference was organized by the new Sonderforschungsbereich 17o "Geometrie und Analysis" at the Mathematical Institute in Göttingen.

H.Grauert

TABLE OF CONTENTS

SFB 17o "Geometrie und Analysis", Conference on Complex Analysis and Algebraic Geometry"

Amar,E. - Wuppertal
Barlet,D. - Nancy
Beauville,A. - Anger
Bohnhorst,G. - Osnabrück
Brieskorn,E. - Bonn
Buchweiz,R.O. - Hannover
Bruns,W. - Vechta
Diederich,K. - Wuppertal
Ebeling,W. - Bonn
v.Essen,H. - Göttingen
Flenner,H. - Göttingen
Fritzsche,K. - Wuppertal
Gerritzen,L. - Bochum
Grauert,H. - Göttingen
Green,M. - Los Angeles
Greuel,G.-M. - Kaiserslautern
Hamm,H. - Münster
Huckleberry,A.T. - Bochum
Hulek,K. - Erlangen
Kahn,C. - Bonn
Karras,U. - Dortmund
Kawamata,Y. - Tokyo
Knörrer,H. - Bonn
Kobayashi,S. - Berkeley
Kohn,J.J. - Princeton
Kosarew,S. - Göttingen
Kriete,H. - Bremen
Kuhlmann,N. - Essen
Kurke,H. - Berlin/DDR
Lieb,I. - Bonn
Lübke,M. - Bayreuth
Maisel,W. - Vechta
Maruyama,M. - Kyoto
Mok,N. - Princeton
Oeljeklaus,E. - Bremen
Okonek,Chr. - Göttingen
Peternell,M. - Wuppertal
Peternell,Th. - Münster (Bayreuth)
Peternell,U. - Münster
Pflug,P. - Vechta
Reiffen,R. - Osnabrück
Remmert,R. - Münster
Richthofer,W. - Bochum
Riemenschneider,O. - Hamburg
Sato,E. - Bonn
Schneider,M. - Bayreuth
Schreyer,H.J. - Kaiserslautern
Schumacher,G. - Münster
Selder,E. - Osnabrück
Shiffman,B. - Baltimore
Siu,Y.T. - Cambridge/Mass.
Skoda,M. - Paris
Sommese,A.J. - Notre Dame
Spindler,H. - Göttingen
Steinbiß,I. - Göttingen
Steinbiß,V. - Göttingen
Stoll,W. - Notre Dame
Timmerscheidt,K. - Essen
Tomassini,G. - Florenz
Trautmann,G. - Kaiserslautern
Van de Ven,A. - Leiden
Vetter,U. - Vechta
Wardelmann,R. - Göttingen

MONODROMY AND POLES OF $\int_X |f|^{2\lambda}\square$

by D. BARLET

The purpose of this talk is, first to give a survey on the results obtained last years on the poles of the meromorphic extension of the distribution $|f|^{2\lambda}$ where f is a non constant holomorphic function on $\mathbb{C}^{n+1}$, and secondly to present recent works in this subject.

§ 1.- Complex Mellin transform and asymptotic expansions

Let X be a connected complex manifold of dimension n+1 and $f : X \to \mathbb{C}$ a non constant holomorphic function. If φ is a C^∞ (n,n) differential form with compact support in X we can consider the continuous function on $\mathbb{C}$ defined by

$$F\varphi(s) = \int_{f=s} \varphi \qquad s \in \mathbb{C} .$$

To study the behaviour of such a function near a critical value of f, it is enough, because the problem is local on X by a simple argument of partition of unity, to consider the case where 0 is the only critical value of f on X. Then the function $F\varphi$ is C^∞ outside 0 and has a bounded support. As is the real case, we shall introduce the (complex) Mellin transform :

$$\mathcal{M}F\varphi(\lambda,m) = \int_{\mathbb{C}} s^{\lambda+m}\, \bar{s}^{\lambda-m}\, F\varphi(s)\, \frac{ds}{s} \wedge \frac{d\bar{s}}{\bar{s}}$$

where $(\lambda,m) \in \mathbb{C} \times (\frac{1}{2}\mathbb{Z})^{(*)}$ satisfies $\mathrm{Re}(\lambda) > 0$ to have a convergent integral. One can now obtain informations about the behaviour of $F\varphi$ when s goes to 0 the study of the meromorphic extension of $\mathcal{M}F\varphi$ on $\mathbb{C} \times (\frac{1}{2}\mathbb{Z})$, its poles and it growth using the following theorem :

Theorem ([B.M.] TH. 4)

Let $(p_j)_{j\in\mathbb{N}}$ and $(q_k)_{k\in\mathbb{N}}$ be sequences of complex numbers such that $(\mathrm{Re}(p_j))_{j\in\mathbb{N}}$ and $(\mathrm{Re}(q_k))_{k\in\mathbb{N}}$ are strictly increasing and unbounded. Let $(n_{j,k})_{(j,k)\in\mathbb{N}^2}$ be non negative integers. The complex Mellin transform is a bijection between functions Φ C^∞ on $\mathbb{C} - \{0\}$, with support in $\{s\ \mathbb{C}/|s| \leq R\}$ and admitting when $s \to 0$ infinitely termwise differentiable expansion

$$\Phi(s) \sim \sum_{\ell\leq n_{j,k}} a_{j,k,\ell}\, s^{p_j}\, \bar{s}^{q_k}\, (\mathrm{Log}\,|s|)^\ell$$

and meromorphic functions M on $\mathbb{C} \times (\frac{1}{2}\mathbb{Z})$ with the following conditions :

1°) the poles of M are of the form $(-p_j-1, -q_k-1)$ with order at most $n_{j,k}+1$ for $(j,k) \in \mathbb{N}^2$

(*) this complex parameter space appears because the complex Mellin transform is some kind of radial (real) Mellin transform and Fourier serie in the angular variable.

2°) $\forall N > 0 \quad \forall \alpha \in \mathbb{N} \quad \exists C(\alpha,N)$ such that

$$|\lambda|^{\alpha} \; |M(\lambda,m)| \leq C(\alpha,N) \exp[2 \operatorname{Re}(\lambda) \operatorname{Log} R]$$

for $\quad \operatorname{Re}(2\lambda+N) > -2$ and $|\operatorname{Im} \lambda| \gg 0$ ■

The interesting case for our purpose is when $(p_j)_{j \in \mathbb{N}}$ and $(q_k)_{k \in \mathbb{N}}$ coincide and are contained in the set $A - \mathbb{N}$ where A is finite and contained in $\mathbb{Q}^-$ (strictly negative rational numbers) and when $n_{j,k}$ is constant (equal to n).

This will allow us to give a new proof of the asymptotic expansion theorem for functions of the type $F\varphi(s) = \int_{f=s} \varphi$ (see [B.1]) if we can prove that the conditions 1°) and 2°) are satisfied for the Mellin transform of $F\varphi$. But for $\operatorname{Re}(\lambda) > 0$ Fubini's theorem gives

$$F\varphi(\lambda,m) = \int_{\mathbb{C}} s^{\lambda+m} \, \bar{s}^{\lambda-m} \left(\int_{f=s} \varphi \right) \frac{ds}{s} \wedge \frac{d\bar{s}}{\bar{s}} = \int_X f^{\lambda+m} \, \bar{f}^{\lambda-m} \, \varphi \wedge \frac{df}{f} \wedge \frac{d\bar{f}}{\bar{f}} \; .$$

So the meromorphic extension of $\mathcal{M} F\varphi$ can be deduced from the existence of Bernstein-Sato polynomial of f (see [$\mathfrak{B}$], [K.1] or [Bj]).

In fact we obtain all the condition 1°) from this result (but without the knowledge of the precise bound n for $n_{j,k}$).

The estimates for condition 2°) are more difficult to prove and we were compelled to use a rather-deep result from M. Kashiwara [K.1] to finish our proof in [B.M.].

§ 2.- Existence of poles for $\int_X |f|^{2\lambda}$ □

Let us consider now a local situation of a representative $f : X \to D$ of a non constant germ of holomorphic function at the origine in $\mathbb{C}^{n+1}$. We shall then assume that 0 is the only critical value of f in X . The "general principle" that we want to illustrate about asymptotic expansions of functions of the type $F\varphi(s) = \int_{f=s} \varphi$ (here φ is a (n,n) C^∞ form with compact support in X as above) is the following :

one can read on these asymptotic expansions the nature of the singularity of the map f .

To give some credibility to this "principle" it is necessary to show that if f has really a critical value at 0, some non trivial term (that is to say some non C^∞ term) will appear in the asymptotic expansion at $s = 0$ of $\int_{f=s} \varphi$ for some well chosen φ. The result we shall give now is in fact more precise : for an isolated singularity it implies clearly our statement above ; in the general case, using the existence of a complex stratification such that the topological type of f is constant along each stratum, result which was proved by Lê Dũng Tráng in [L], one can reduce to the isolated singularity case : cutting transversally the biggest stratum contained in the singular locus of $\{f = 0\}$ one can exhibit a point of

$\{f = 0\}$ where the Milnor's fiber of f is not homologically of dimension 0 ; then by theorem 2 and theorem 3 we can conclude that some non C^∞ term will appear in the asymptotic expansion for some suitable choice of φ .

Now let us assume that $f : X \to D$ is a Milnor representative of the chosen germ, that is to say that

1°) X is a Stein an contractible open set

2°) $f|X-f^{-1}(0) \to D - \{0\}$ is a C^∞ locally trivial fibration whose fiber has finite dimensional cohomology.

In fact Milnor has proved (see [$\mathcal{M}$]) that such conditions are fullfilled for a basis of neighbourhood of 0 in $\mathbb{C}^{n+1}$ and that the C^∞ fibration does not depend of the choices. In this situation one can define a monodromy operator

$$T : H^*(f^{-1}(s_0),\mathbb{C}) \to H^*(f^{-1}(s_0),\mathbb{C})$$

where s_0 is a base point in $D - \{0\}$:

for $\Delta \subset D - \{0\}$ a disc such that $f|f^{-1}(\Delta) \to \Delta$ is C^∞ isomorphic to $\Delta \times f^{-1}(s_1)$ for $s_1 \in \Delta$, the canonical map

$$H^*(f^{-1}(\Delta),\mathbb{C}) \to H^*(f^{-1}(s_1),\mathbb{C})$$

is an isomorphism. This leads, for s_1 and $s_2 \in \Delta$ to a <u>canonical</u> isomorphism

$$\Phi^{\Delta}_{s_1,s_2} : H^*(f^{-1}(s_1),\mathbb{C}) \to H^*(f^{-1}(s_2),\mathbb{C}) \quad .$$

We define a locally constant family of cohomology classes $c_s \in H^p(f^{-1}(s),\mathbb{C})$ for s in a subset $S \subset D$ iff for $s_1,s_2 \in \Delta \cap S$ we have

$$\Phi^{\Delta}_{s_1,s_2}(c_{s_1}) = c_{s_2}$$

when Δ is a small disc as above.

It is then clear that a locally constant family of cohomology classes on a disc $\Delta \subset D - \{0\}$ is determined by its value at any point of Δ . If $\gamma \in H^p(f^{-1}(s_0),\mathbb{C})$ one can follow a locally constant family of cohomology classes along a closed counter-clockwise path going from s_0 to s_0 $(t \to e^{it} s_0 \quad t \in [0,2\pi])$.
Then put $T^p(\gamma)$ be the value of this locally constant family at $t = 2\pi$. For instance, let $w \in \Omega^p(X)$ an holomorphic p-form satisfying $dw = (m+u)\frac{df}{f} \wedge w$ on $X - f^{-1}(0)$ where $m \in \mathbb{N}$ and $0 \leq \mathrm{Re}(u) < 1$. Then for $\Delta \subset D - \{0\}$ a disc, we can choose a determination of Log f on $f^{-1}(\Delta)$ and consider $\frac{w}{f^{m+u}}$ as a closed holomorphic p-form on the Stein manifold $f^{-1}(\Delta)$.
Then $\frac{w}{f^{m+u}}$ induces a class in $H^p(f^{-1}(\Delta),\mathbb{C})$ and the family $s \to \frac{w}{f^{m+u}}\Big|_{f^{-1}(s)} \in H^p(f^{-1}(s),\mathbb{C})$ is (locally) constant on Δ . Starting from s_0 what will happen when we compute $T(\gamma)$ for $\gamma \in H^p(f^{-1}(s_0),\mathbb{C})$ induced by $\frac{w}{f^{m+u}}\Big|_{f^{-1}(s_0)}$ is clear : nothing happens to w since w is uniform on X. So we just have

$T\gamma = e^{-2i\pi u} \cdot \gamma$.

In fact it is possible to prove (but not obvious) that any eigenvector of the monodromy can be built in this way (see [B.2] lemme A) and this is the starting point of the proof of the

Theorem 2 ([B.2])

Let $f : X \to D$ be a Milnor respresentative of a non constant germ at 0 of holomorphic function in $\mathbb{C}^{n+1}$. Assume that $e^{-2i\pi u}$ is eigenvalue of multiplicity k of the monodromy acting on the cohomology of the Milnor's fiber of f. Then for $\nu \in \mathbb{N}$ large enough the meromorphic extension of $\int_X |f|^{2\lambda} \square$ has a pole of order at least k at the point $-u-\nu$ ∎

One application of this result is to prove, using the following theorem (Malgrange [M], Kashiwara [K.2] ...).

Theorem

If α is a root of the Bernstein-Sato polynomial of f at 0, then there exist $x_0 \in f^{-1}(0)$ such that monodromy acting on the cohomology of Milnor's fiber of f at x_0 admits $e^{2i\pi\alpha}$ as eigenvalue ∎

the following corollary :

Corollary

To any root α of the Bernstein-Sato polynomial of f at 0 correspond poles of the meromorphic extension of $\int_X |f|^{2\lambda} \square$ at $\alpha - \nu$ for $\nu \in \mathbb{N}$ large enough ∎

Of course, from theorem 2, one can also deduces more informations about the Bernstein-Sato polynomial of f at 0, denote it by b, since

$$b(\lambda)\, b(\lambda+1) \ldots b(\lambda+N) \int_X |f|^{2\lambda} \square$$

has no poles for $\mathrm{Re}(2\lambda+N) > 0 \quad \forall N \in \mathbb{N}$ using Bernstein identity (see [B.2] lemme 1 for instance).

Going back to our "general principle" we see that the theorem 2 is not enough in the case of the eigenvalue 1 of the monodromy (acting in strictly positive degree) : even if f is smooth (i.e. $df \neq 0$ on X) the meromorphic extension of $\int_X |f|^{2\lambda} \square$ has simple poles at negative integers (just because the Mellin transform of C^∞ functions has such poles ; we assume here $f(0) = 0$). So if the monodromy has 1 as simple eigenvalue when acting in positive degree on the cohomology of the Milnor fiber, this cannot be seen via the theorem 2. This gap is filled up by theorem 3 which is in fact a special case of a theorem which shows that if eigenvectors for a same eigenvalue of the monodromy (but may be acting in different degrees) have a non trivial cup-product relation, this forces the corresponding poles to be of higher order (for a precise statement of the general theorem see [B.3]).

Theorem 3 ([B.3] corollary 1)

Let $f : X \to D$ be a Milnor representative of a non constant germ at 0 of holomorphic function in $\mathbb{C}^{n+1}$. Assume that 1 is eigenvalue of multiplicity k of the monodromy acting in strictly positive degree on the cohomology of the Milnor's fiber of f. Then for $\nu \in \mathbb{N}$ large enough, $\int_X |f|^{2\lambda} \square$ has a pole of order at least k+1 at the point $-\nu$ ■

§ 3.- Where does the pole appear ?

We want now to explain some recent paper where we try to give an answer to the question "where does the pole appear" ? The main result is the

Theorem 4

Let $f : X \to D$ be a Milnor representative of a non constant germ at 0 of holomorphic function in $\mathbb{C}^{n+1}$. Assume that $e^{-2i\pi u}$ is eigenvalue of multiplicity k for the monodromy acting on $H^p(f^{-1}(s_0),\mathbb{C})$, where $0 \leq u < 1$ (s_0 is a base point in $D - \{0\}$ so $f^{-1}(s_0)$ is Milnor's fiber). Then the meromorphic extension of $\int_X |f|^{2\lambda}\square$ has a pole of order at least k at $-p-u$ ■

It is not hard to check that in general $-p-u$ is the best possible : take n=1 and $f(x,y) = x^2 - y^3$ then $\omega = 3x\,dy - 2y\,dx$ satisfies

$$d(y\omega) = \frac{7}{6}\frac{df}{f} \wedge (y\omega) \quad \text{on} \quad \mathbb{C}^2$$

and (as explained above) it gives an eigenvector for the monodromy acting on $H^1(f^{-1}(1),\mathbb{C})$ with eigenvalue $e^{-2i\pi\, 7/6}$ (because quasi-homogeneity we have a global Milnor representative $f : \mathbb{C}^2 \to \mathbb{C}$). The first pole of $\int_X |f|^{2\lambda} \square$ which is equal modulo $\mathbb{Z}$ to $-\frac{1}{6}$ is $-\frac{7}{6}$ in this case (a detailed computation of asymptotic expansion for this example can be found in [E]).

We shall try to explain the main idea in the proof of theorem 4.

If V is a Stein manifold of dimension n, and if you take an element e in $H^p(V,\mathbb{C})$ you can represent this cohomology class by an holomorphic d-closed p-form w on V. Now because $H^p(V,\mathbb{C})$ has a natural complex conjuguation (coming from the isomorphism $H^p(V,\mathbb{C}) = H^p(V,\mathbb{R}) \otimes_{\mathbb{R}} \mathbb{C}$) we can try to find an holomorphic d-closed p-form on V representing the cohomology class $\bar{e}$, that is to say cohomologous to $\bar{w}$. The answer is given by chasing in the diagram given by the bigraduate De Rham complex of V. Explicitely we do the following :

w is a ∂-closed p-form on V. For $p \geq 1$ Cartan's theorem B (conjuguate) gives a C^∞(p-1) form of type (p-1,0), say $v^{p-1,0}$, such that $\partial v^{p-1,0} = w$ on V. Now $\bar{\partial} v^{p-1,0}$ is a ∂-closed form of type (p-1,1) and if $p \geq 2$ we can continue. Let us just finish the case p = 1 : as $\bar{\partial} v^{0,0}$ is of type (0,1) and ∂-closed this is an antiholomorphic form on V ; call ω its conjuguate. We have then $w = d\, v^{0,0} - \bar{\omega}$ on V and this shows that $-\omega$ is cohomologous to $\bar{w}$.

Now in the case of Milnor's fibration we want to play the same kind of game with a parameter ; we consider a relatively closed form on the family on the Stein manifolds $f^{-1}(s)$ for $s \in D - \{0\}$. We begin with a form $w \in \Omega^p(X)$ satisfying on X

$$dw = (m+u)\,\frac{df}{f} \wedge w \qquad m \in \mathbb{N} \quad 0 \leq \operatorname{Re} u < 1$$

(so $\frac{w}{f^{m+u}}$ induces a multivalued locally constant family of cohomology classes corresponding to an eigenvector of the monodromy for the eigenvalue $e^{-2i\pi u}$).
To enter the game we have to define a ∂-closed current on $X^{(*)}$. This is possible if we can define $\frac{w}{|f|^{2(m+u)}}$ as a ∂-closed current on X. We try to do it by tacking the value at $\lambda = -m-u$ of the meromorphic extension of the current $\int_X |f|^{2\lambda}\, w \wedge \square$; but there is an obstruction to the ∂-closedness : if Pf denotes the constant term in the Laurent expansion we have (exercice)

$$d'(\operatorname{Pf}(\lambda = -m-u, \int_X |f|^{2\lambda}\, w \wedge \square) = \pm \operatorname{Res}(\lambda = -m-u, \int_X |f|^{2\lambda}\, \frac{df}{f} \wedge w \wedge \square\,)$$

as currents on X (Res is the residu). So our game will work if we have no poles at $\lambda = -m-u$ for the meromorphic extension of $\int_X |f|^{2\lambda}\, \frac{df}{f} \wedge w \wedge \square$.
This is roughly the main idea of the fundamental proposition in [B.4]. The above description suggests that, although the Hodge structure of a Stein manifold is trivial, some degeneration of the variation of this Hodge structure in a family with a singular fiber yelds to non trivial phenomena. Thinking to Varchenko's description of Steenbrink's mixed Hodge structure$^{(**)}$ in the isolated singularity case, this is not so surprising. But in the non isolated singularity case it is not possible, in general, to algebrize the situation.
The theorem 4 leads easily to the following information on the Bernstein-Sato polynomial of f.

Corollary

Let $f : X \to D$ as in theorem 4 and assume that the monodromy acting on $H^p(f^{-1}(s_0),\mathbb{C})$, the p-th cohomology group of Milnor's fiber, has $e^{-2i\pi u}$ as eigenvalue of multiplicity k with $0 \leq u < 1$. Then $b^{(***)}$ has at least k roots (counting multiplicities) of the form $-q-u$ for $q \in \mathbb{N}$, $0 \leq q \leq p$ ∎

(*) inducing on $f = s$ almost the same cohomology class than $\frac{w}{f^{m+u}}$ as we shall see.

(**) see [S] and [V].

(***) b denotes the Bernstein-Sato polynomial of f at 0.

References

[B.1] D. Barlet, Développements asymptotiques des fonctions obtenues par intégration dans les fibres, Inv. Math. 68 (1982), p. 129-174.

[B.2] D. Barlet, Contribution effective de la monodromie aux développements asymptotiques, Ann. Scient. Ec. Norm. Sup. 4ème série, t. 17 (1984), p. 293-315.

[B.3] D. Barlet, Contribution du cup produit de la fibre de Milnor aux pôles de $|f|^{2\lambda}$, Ann. Inst. Fourier (Grenoble) t. 34, fasc. 4 (1984), p. 75-107.

[B.4] D. Barlet, Monodromie et pôles de $\int_X |f|^{2\lambda} \square$, preprint Nancy Avril 84, à paraître au Bull. Soc. Math. France.

[B.M.] D. Barlet et H.M. Maire, Développements asymptotiques, transformation de Mellin complexe et intégration sur les fibres, preprint déc. 84 (Nancy/Genève).

[$\mathfrak{B}$] J.N. Bernstein, Prolongement analytique des fonctions généralisées avec paramètres (en russe) Funkts-Analys 6,4 (1972) p. 26-40.

[Bj] J.E. Björk, Rings of differential operators, North Holland (1979).

[E] M. El Amrani, Singularités des fonctions obtenues par intégration sur la fibre $X^2 - Y^3 = s$ et identités modulaires, Bull. Sci. Math. 2ème série 108 (1984)

[K.1] M. Kashiwara, b-functions and holonomic systems, rationnality of roots of b-functions, Inv. Math. 38 (1976) p. 33-53.

[K.2] M. Kashiwara, Holonomic systems of linear differential equations with regular singularities and related topics in topology, Advanced Studies in Pure Math. 1 1983 , Algebraic Varieties and Analytic Varieties p. 49-54.

[L] Lê Dũng Tráng, Some remarks on relative monodromy in Nordic Summer School Symposium in Math. Oslo 1976.

[M] B. Malgrange, Polynômes de Bernstein-Sato et cohomologie évanescente, Astérisque 101 (1982) Soc. Math. France.

[$\mathcal{M}$] J. Milnor, Singular points of complex hypersurfaces, Annals of Math. Studies 61 Princeton University Press 1968.

[S] J. Steenbrink, Mixed Hodge structure on the vanishing cohomology in Nordic Summer School Symposium in Math. Oslo 1976.

[V] A. Varchenko, Mixed Hodge structure on the vanishing cohomology, Izv. Akad. Nauk SSSR sci. math. 45,3 (1981) p. 540-591.

D. Barlet - Institut Elie Cartan
U.A. 750 - Faculté des Sciences
Université de Nancy I - B.P. 239
54506 - Vandœuvre les Nancy Cedex

LE GROUPE DE MONODROMIE DES FAMILLES UNIVERSELLES D'HYPERSURFACES ET D'INTERSECTIONS COMPLETES

Arnaud BEAUVILLE

Le but de cet exposé est d'illustrer deux beaux théorèmes de Janssen [J] et Ebeling [E]. Ces résultats, de nature purement algébrique, permettent à leurs auteurs de calculer, dans un grand nombre de cas, la monodromie des singularités isolées. Je voudrais montrer sur un exemple qu'ils s'appliquent également très bien au calcul de la monodromie des familles de variétés lisses.

Le problème précis que je veux traiter est le suivant. Les hypersurfaces de degré d dans $\mathbb{P}^{n+1}$ sont paramétrées par un espace projectif $\mathbb{P}^{N(n,d)}$ (avec $N(n,d) = \binom{n+d+1}{d}-1$), dans lequel les hypersurfaces lisses forment un ouvert $U_{n,d}$. Soient u un point de $U_{n,d}$, et X l'hypersurface correspondante. Le groupe $\pi_1(U_{n,d},u)$ opère sur $H^n(X,\mathbb{Z})$, et l'image $\Gamma_{n,d}$ de l'homomorphisme $\rho : \pi_1(U_{n,d},u) \longrightarrow \mathrm{Aut}(H^n(X,\mathbb{Z}))$ est appelée le <u>groupe de monodromie</u> de la famille universelle des hypersurfaces de dimension n et de degré d; c'est ce groupe que nous allons déterminer. Après quelques préliminaires (n° 1), le résultat est obtenu au n° 2 (n pair) et au n° 3 (n impair). Quelques applications sont données au n° 4; on condidère au n° 5 la situation plus générale des intersections complètes.

1. Réseaux évanescents.

Soit L un $\mathbb{Z}$-module libre de type fini, muni d'une forme bilinéaire symétrique ou alternée, notée $(a,b) \longmapsto a.b$. Soit Δ un ensemble d'éléments de L; si la forme est symétrique, nous supposons $\delta^2 = \pm 2$ pour tout $\delta \in \Delta$. Soit s_δ l'automorphisme de L défini par $s_\delta(x) = x + (\delta.x)\delta$ si $\delta^2 = 0$ ou -2, $s_\delta(x) = x - (\delta.x)\delta$ si $\delta^2 = 2$; il respecte la forme bilinéaire. On note Γ_Δ le groupe d'automorphismes de L engendré par les s_δ pour $\delta \in \Delta$.

DEFINITION.- <u>On dit que le couple</u> (L,Δ) <u>est un réseau évanescent si</u> Δ <u>engendre</u> L, <u>et si</u> Δ <u>est une orbite de</u> Γ_Δ <u>dans</u> L.

Indiquons deux exemples de réseaux évanescents, en renvoyant pour les détails aux articles [E] et [J].

Soit F la fibre de Milnor d'une singularité isolée de dimension n; le <u>module des cycles évanescents</u> $H_n(F,\mathbb{Z})$, muni de la forme d'intersection et de l'ensemble Δ_F des cycles évanescents, est un réseau évanescent : c'est ce qui explique à la fois la terminologie et l'intérêt de cette notion pour l'étude des singularités.

Voici un autre exemple, fondamental pour ce qui suit. Soit $(X_t)_{t\in\mathbb{P}^1}$ un pinceau de Lefschetz d'hypersurfaces de degré d dans $\mathbb{P}^{n+1}$; notons S l'ensemble (fini) des points $s\in\mathbb{P}^1$ tels que X_s soit singulière. Fixons $t\in\mathbb{P}^1-S$, et posons $X=X_t$. Notons $H^n(X,\mathbb{Z})_o$ la cohomologie primitive (égale à $H^n(X,\mathbb{Z})$ si n est impair, et à l'orthogonal de $h^{n/2}$ dans $H^n(X,\mathbb{Z})$ si n est pair). Pour chaque $s\in S$, le choix d'un chemin dans $\mathbb{P}^1-S$ joignant t à un point voisin de s détermine un cycle évanescent $\delta\in H^n(X,\mathbb{Z})$; notons Δ_X l'ensemble de ces cycles évanescents. D'après la théorie de Lefschetz, le couple $(H^n(X,\mathbb{Z})_o,\Delta_X)$ est un réseau évanescent.

Si le pinceau choisi est assez général, un théorème de Zariski entraîne que l'inclusion $\mathbb{P}^1-S\subset U_{n,d}$ induit un homomorphisme surjectif sur les π_1; le groupe de monodromie $\Gamma_{n,d}$ coïncide donc avec le groupe Γ_Δ du réseau évanescent $(H^n(X,\mathbb{Z})_o,\Delta_X)$.

Indiquons maintenant une relation entre les deux constructions précédentes. Soit X_o une hypersurface de degré d dans $\mathbb{P}^{n+1}$, admettant une singularité isolée en un point p. La fibre de Milnor F de cette singularité est l'intersection d'une hypersurface X_ε lisse voisine de X_o avec une petite boule de centre p dans $\mathbb{P}^{n+1}$. Par transport de X_ε à X on en déduit un homomorphisme $i : H_n(F,\mathbb{Z}) \longrightarrow H_n(X,\mathbb{Z}) = H^n(X,\mathbb{Z})$, qui est compatible aux formes d'intersection et applique Δ_F dans Δ_X. Si la forme d'intersection sur $H_n(F,\mathbb{Z})$ est non dégénérée, i est nécessairement injective, et identifie donc $(H_n(F,\mathbb{Z}),\Delta_F)$ à un sous-réseau évanescent de $(H^n(X,\mathbb{Z})_o,\Delta_X)$.

2. Dimension paire.

Soit M un réseau orthogonal. Si $M_{\mathbb{R}}$ est non dégénéré, il existe pour $\varepsilon=\pm1$ un homomorphisme $\sigma_\varepsilon : O(M_{\mathbb{R}}) \longrightarrow \{\pm1\}$ caractérisé par la propriété suivante : si v est un vecteur non isotrope de $M_{\mathbb{R}}$ et si s_v désigne la réflection orthogonale par rapport à $v^\perp$, on a $\sigma_\varepsilon(s_v)=\varepsilon v^2/|v^2|$. Si $M_{\mathbb{R}}$ est dégénéré, de noyau K, on note encore σ_ε le composé $O(M_{\mathbb{R}}) \longrightarrow O(M_{\mathbb{R}}/K) \xrightarrow{\sigma_\varepsilon} \{\pm1\}$. D'autre part, on note $D(M)$ le conoyau de l'homomorphisme $M \longrightarrow M^*$ associé à la forme bilinéaire, et τ l'homomorphisme canonique $O(M) \longrightarrow \mathrm{Aut}\,(D(M))$.

Soit (L,Δ) un réseau évanescent orthogonal. On pose $O^*(L)=O(L)\cap\operatorname{Ker}\sigma_\varepsilon\cap\operatorname{Ker}\tau$, avec $\varepsilon=\frac{1}{2}\delta^2$ pour $\delta\in\Delta$. Il est clair que Γ_Δ est contenu dans $O^*(L)$. Par ailleurs, on dit comme dans [E] que le réseau évanescent (L,Δ) est <u>complet</u> si Δ contient 6 éléments dont le diagramme d'intersection est

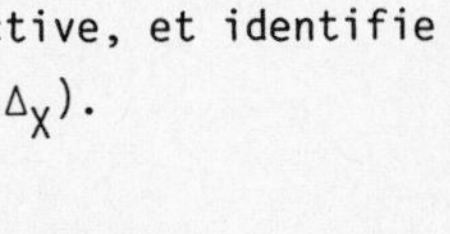

(comme d'habitude, les éléments correspondent aux sommets du diagramme, le produit de deux éléments étant égal au nombre de traits joignant les sommets correspondants).

THEOREME 1 [E].- <u>Soit</u> (L,Δ) <u>un réseau évanescent orthogonal complet</u>. <u>On a</u> $\Gamma_\Delta = O^*(L)$.

Nous allons en déduire le groupe $\Gamma_{n,d}$ pour n pair. Il est clair que $\Gamma_{n,d}$ est contenu dans le sous-groupe $O_h(H^n(X,\mathbb{Z}))$ de $O(H^n(X,\mathbb{Z}))$ formé des automorphismes qui préservent $h^{n/2}$ (on désigne par h la classe dans $H^2(X,\mathbb{Z})$ d'une section hyperplane). Posons $O_h^+(H^n(X,\mathbb{Z})) = O_h(H^n(X,\mathbb{Z})) \cap \operatorname{Ker} \sigma_\varepsilon$, avec $\varepsilon = (-1)^{n/2}$.

THEOREME 2.- <u>Pour</u> n <u>pair, le groupe de monodromie</u> $\Gamma_{n,d}$ <u>est le groupe</u> $O_h^+(H^n(X,\mathbb{Z}))$. <u>Il est d'indice</u> 2 <u>dans</u> $O_h(H^n(X,\mathbb{Z}))$ <u>si</u> $d \geq 4$ <u>ou</u> $d = 3$ <u>et</u> $n \neq 2$, <u>d'indice</u> 1 <u>dans les autres cas</u>.

Le cas des quadriques est laissé au lecteur en exercice. Le cas des surfaces cubiques est bien connu : le groupe de monodromie $\Gamma_{2,3}$ est le groupe de Weyl $W(E_6)$, égal à $O_h(H^2(X,\mathbb{Z}))$. Supposons désormais $d \geq 3$, et $n \geq 4$ si $d = 3$. Nous allons montrer que le réseau évanescent $(H^n(X,\mathbb{Z})_o, \Delta_X)$ est complet. Nous utiliserons pour cela une des singularités exceptionnelles d'Arnold [A], la singularité U_{12} : pour une surface, elle est donnée analytiquement dans $\mathbb{C}^3$ par l'équation $x^3 + y^3 + z^4 = 0$. D'après [E], 5.3, le réseau évanescent de cette singularité est complet. Il en est donc de même de tout réseau évanescent le contenant; compte tenu des remarques du n° 1, il suffit donc de prouver qu'il existe une hypersurface de degré d dans $\mathbb{P}^{n+1}$ admettant une singularité de type U_{12}. Pour $d \geq 4$, il suffit de prendre l'hypersurface d'équation affine

$$x_1^3 + x_2^3 + x_3^4 + x_3^d + \sum_{i=4}^{n+1} x_i^2 = 0 .$$

Pour $d = 3$, on remarque qu'il existe une surface cubique, d'équation affine $f(x_1,x_2,x_3) = 0$ dans $\mathbb{C}^3$, admettant à l'origine une singularité de type E_6 (c'est-à-dire d'équation analytique $z^2 + x^3 + y^4 = 0$; c'est un résultat classique, cf. par exemple [D]). L'hypersurface d'équation affine

$$f(x_1,x_2,x_3) + x_4^3 + \sum_{i=5}^{n+1} x_i^2 = 0$$

fait l'affaire, d'où notre assertion.

Considérons maintenant l'homomorphisme de restriction $r : O_h^+(H^n(X,\mathbb{Z})) \longrightarrow O(H^n(X,\mathbb{Z})_o)$; il est injectif, et son image est contenue dans O^*. Comme $r(\Gamma_{n,d}) = O^*$ d'après le Théorème 1, on a $\Gamma_{n,d} = O_h^+(H^n(X,\mathbb{Z}))$. Enfin le réseau complet $H^n(X,\mathbb{Z})_o$ contient un élément η de carré -2ε; on a $\sigma_\varepsilon(s_n) = -1$, de sorte que O_h^+ est d'indice 2 dans O_h.

3. <u>Dimension impaire</u>.

Soit (L,Δ) un réseau évanescent symplectique. Le groupe Γ_Δ est alors contenu dans le groupe symplectique $Sp(L)$. On appelle <u>forme quadratique mod. 2</u> sur le réseau symplectique L une fonction $q : L \longrightarrow \mathbb{Z}/2\mathbb{Z}$ satisfaisant à $q(x+y) = q(x) + q(y) + (x.y)$. Si on a $q(\delta) = 1$ pour tout $\delta \in \Delta$, le groupe Γ_Δ est contenu dans le sous-groupe $SpO(L,q)$ de $Sp(L)$ formé des automorphismes symplectiques qui préservent q. Inversement, si la transvection s_δ préserve q et est distincte de l'identité, on a $q(\delta) = 1$.

La classification de Janssen est relativement compliquée; je vais me contenter d'énoncer un résultat beaucoup plus faible (que l'on déduit très facilement de [J]).

THEOREME 3.- _Soit_ (L,Δ) _un réseau évanescent symplectique unimodulaire. Supposons que_ Δ _contienne_ 6 _éléments dont le diagramme d'intersection_ mod. 2 _est de type_ E_6. _On a alors l'une des deux possibilités suivantes_ :

(i) $\Gamma_\Delta = Sp(L)$;

(ii) _il existe une forme quadratique_ mod. 2 q _sur_ L _telle que_ $\Gamma_\Delta = SpO(L,q)$.

THEOREME 4.- _Supposons_ n _impair_.

(i) _Si_ d _est pair, le groupe de monodromie_ $\Gamma_{n,d}$ _est le groupe symplectique_ $Sp(H^n(X,\mathbb{Z}))$;

(ii) _si_ d _est impair, il existe sur_ $H^n(X,\mathbb{Z})$ _une forme quadratique_ mod. 2 q_X _invariante par monodromie, et on a_ $\Gamma_{n,d} = SpO(H^n(X,\mathbb{Z}),q_X)$.

On laisse de nouveau au lecteur le cas des quadriques, ainsi que celui des cubiques planes. Montrons dans les autres cas que les hypothèses du Théorème 3 sont satisfaites. Il suffit comme ci-dessus de mettre en évidence une hypersurface (de dimension n et de degré d) admettant une singularité de type E_6. Pour $d \geq 4$, on prend l'hypersurface d'équation affine

$$x_1^3 + x_2^4 + x_2^d + \sum_{i=3}^{n+1} x_i^2 = 0 .$$

Pour $d = 3$, on utilise encore la surface cubique avec une singularité de type E_6, en ajoutant une somme de carrés à son équation.

On est donc dans l'une des situations (i) ou (ii) du Théorème 3. Si d est impair, il existe effectivement une forme quadratique mod. 2 sur $H^n(X,\mathbb{Z})$ invariante par déformation (cf. [B] ou [W], ou aussi n° 4c ci-dessous), de sorte qu'on est dans le cas (ii).

Il reste à montrer que lorsque d est pair (≥ 4), il n'existe pas de forme quadratique mod. 2 q sur $H^n(X,\mathbb{Z})$ invariante par monodromie, c'est-à-dire telle que $q(\delta) = 1$ pour tout $\delta \in \Delta_X$. Il suffit pour cela de trouver un nombre _impair_ d'éléments $\delta_1,\ldots,\delta_{2p+1}$ de Δ_X, deux à deux orthogonaux, vérifiant $\Sigma\, \delta_i \equiv 0$ (mod. 2). Pour réaliser cette situation, nous allons considérer un pinceau d'hypersurfaces

$(X_t)_{t\in\mathbb{P}^1}$, où pour $t=0$ la fibre X_o acquière un certain nombre de points doubles ordinaires $P_1,\dots,P_\ell$. A chaque P_i est associé un cycle évanescent δ_i dans l'homologie d'une fibre voisine X (c'est-à-dire, par dualité de Poincaré, dans $H^n(X,\mathbb{Z})$). Supposons qu'il existe une sous-variété lisse Z de X_o, de dimension $(n+1)/2$, passant par les P_i. Soit $B(P_i)$ une petite boule de centre P_i dans Z; en poussant $Z-\bigcup_i B(P_i)$ dans la fibre voisine X, on obtient une $(n+1)$-chaîne dont le bord est $\Sigma\pm\delta_i$. On a donc $\sum_{i=1}^{\ell}\pm\delta_i=0$ dans $H^n(X,\mathbb{Z})$, et les δ_i sont bien sûr deux à deux orthogonaux. Il nous suffit donc de réaliser une telle situation $Z\subset X_o$, avec ℓ impair.

Ecrivons $n=2\nu-1$. Soient $G_1(T_o,\dots,T_\nu),\dots,G_\nu(T_o,\dots,T_\nu)$ des polynômes homogènes de degré $d-1$, supposés assez généraux pour que les hypersurfaces $G_1=0,\dots,G_\nu=0$ dans $\mathbb{P}^\nu$ se rencontrent transversalement en $(d-1)^\nu$ points distincts. Prenons pour X_o l'hypersurface dans $\mathbb{P}^{2\nu}$ d'équation

$$T_{\nu+1}\,G_1(T_o,\dots,T_\nu)+\dots+T_{2\nu}\,G_\nu(T_o,\dots,T_\nu)=0 .$$

On vérifie aussitôt que les seules singularités de X_o sont les $(d-1)^\nu$ points définis par

$$T_{\nu+1}=\dots=T_{2\nu}=G_1=\dots=G_\nu=0 ,$$

et que ce sont des points doubles ordinaires. Ils sont en nombre impair, et sont contenus dans le sous-espace linéaire Z de X_o défini par $T_{\nu+1}=\dots=T_{2\nu}=0$. On a donc obtenu la situation cherchée, ce qui achève la démonstration du théorème.

4. Applications.

a) Hypersurfaces marquées.

Indiquons d'abord une autre formulation des Théorèmes 2 et 4, qui est utile pour les questions de modules et de périodes. Supposons d'abord n pair. Soient X_o une hypersurface lisse de dimension n et de degré d, $h_o\in H^2(X_o,\mathbb{Z})$ la classe d'une section hyperplane. Posons $L_{n,d}=H^n(X_o,\mathbb{Z})$ et $h_{n,d}=h_o^{n/2}$. Le Théorème 2 se traduit comme suit :

PROPOSITION 1.- <u>L'espace des hypersurfaces</u> X <u>de dimension paire</u> n <u>et de degré</u> d, <u>munies d'une isométrie</u> $\phi : L_{n,d}\longrightarrow H^n(X,\mathbb{Z})$ <u>telle que</u> $\phi(h_{n,d})=h^{n/2}$, <u>a deux composantes irréductibles pour</u> $d\geq 4$ <u>ou</u> $d=3$, $n\geq 4$. <u>Il est irréductible dans les autres cas</u>.

Pour traiter le cas n impair, rappelons quelques définitions. Soit L un réseau symplectique unimodulaire. On appelle <u>base symplectique</u> de L une base $e_1,\dots,e_{2g}$ satisfaisant à $e_i\cdot e_j=0$ si $|i-j|\neq g$, $e_i\cdot e_{g+i}=1$ pour $1\leq i\leq g$.

Soit q une forme quadratique mod. 2 sur L. On dit qu'une base symplectique $e_1,\dots,e_{2g}$ est adaptée à q si l'on a $q(e_i)=0$ pour $i \leq 2g-2$ et $q(e_{2g-1})=1$ (on voit facilement qu'une telle base existe toujours). La valeur de $q(e_{2g})$ est alors indépendante du choix de la base; on l'appelle l'invariant d'Arf de q. Le Théorème 4 se traduit ainsi :

PROPOSITION 2.- *L'espace des hypersurfaces X de dimension impaire n et de degré d, munies d'une base symplectique de $H^n(X,\mathbb{Z})$ si d est pair (resp. d'une base symplectique adaptée à q_X, si d est impair) est irréductible.*

Plus généralement les Théorème 2 et 4 permettent de déterminer les composantes irréductibles des espaces de modules d'hypersurfaces munies d'une structure de niveau, c'est-à-dire d'une structure supplémentaire (base ou élément mod p, etc.) sur $H^n(X,\mathbb{Z})$. Je vais me contenter d'un exemple assez classique, celui des thêta-caractéristiques sur les courbes planes.

b) Thêta-caractéristiques.

Soit C une courbe (lisse, compacte) de genre g. Rappelons qu'une thêta-caractéristique sur C peut être définie de l'une des trois manières (équivalentes) suivantes :

(i) un faisceau inversible L sur C (à isomorphisme près), vérifiant $L^{\otimes 2} \simeq \omega_C$;

(ii) un diviseur thêta Θ symétrique sur $J(C)$;

(iii) une forme quadratique q mod. 2 sur $H^1(C,\mathbb{Z})$.

On a $$h^0(L) = \mathrm{mult}_0(\Theta) \equiv \mathrm{Arf}(q) \pmod{2} ;$$

on dit que la thêta-caractéristique est paire (resp. impaire) si ces nombres sont égaux à 0 (resp. 1) dans $\mathbb{Z}/2$.

Les courbes planes de degré d impair admettent une thêta-caractéristique canonique, définie par le faisceau $\mathcal{O}_C(\frac{d-3}{2})$ (ou par la forme q_C).

PROPOSITION 3.- *L'espace des courbes planes de degré $d \geq 4$, munies d'une thêta-caractéristique, admet*

- *deux composantes irréductibles si d est pair, correspondant aux thêta-caractéristiques paires et impaires;*
- *trois composantes si d est impair, correspondant à la thêta-caractéristique canonique et aux autres thêta-caractéristiques paires et impaires.*

Ce résultat a été obtenu auparavant, par une méthode différente, par F. Catanese (non publié).

Soit u un point de $U_{1,d}$, correspondant à la courbe plane C; désignons par L le réseau symplectique $H^1(C,\mathbb{Z})$ et par $Q(L)$ l'ensemble des formes quadratiques mod.2 sur L. Par construction, l'espace que nous considérons est le revêtement étale de $U_{1,d}$ défini par l'action de $\pi_1(U_{1,d},u)$ sur $Q(L)$ (ou, si l'on préfère, le quotient de cet espace par $PGL(3)$; cela ne change rien aux questions d'irréductibilité). Il s'agit donc de décrire les orbites dans $Q(L)$ des groupes $Sp(L)$ et $SpO(L,q)$, pour $q \in Q(L)$.

Il résulte immédiatement de l'existence des bases adaptées que $Q(L)$ est réunion de deux orbites sous $Sp(L)$, distinguées par l'invariant d'Arf. D'autre part, $Q(L)$ est un espace affine sous L : pour $q \in Q(L)$, $x \in L$, la forme $q+x$ est définie par $(q+x)(y) = q(y) + (x.y)$. Le choix d'une origine $q \in Q(L)$ permet d'identifier $Q(L)$ à L, et ce de manière compatible avec l'action de $SpO(L,q)$. Compte tenu de la formule $\mathrm{Arf}\,(q+x) = \mathrm{Arf}\,(q) + q(x)$, on en déduit que $Q(L) - \{q\}$ est réunion de deux orbites sous $SpO(L,q)$, distinguées par l'invariant d'Arf. La proposition en résulte aussitôt.

c) Difféomorphismes.

Soit X une hypersurface lisse de dimension n et de degré d. Notons $\mathrm{Diff}^+(X)$ le groupe des difféomorphismes de X préservant l'orientation (dans la suite, nous dirons simplement difféomorphisme). Ce groupe opère sur la cohomologie en respectant la forme d'intersection, d'où un homomorphisme
$\pi : \mathrm{Diff}^+(X) \longrightarrow \mathrm{Aut}(H^n(X,\mathbb{Z}))$. Puisque $\mathrm{Diff}^+(X)$ contient les difféomorphismes de monodromie, l'image de π contient $\Gamma_{n,d}$. Nous allons déterminer précisément cette image.

PROPOSITION 4.- *Supposons* n *pair* ≥ 4. *L'image de* π *est égale à* $O_h(H^n(X,\mathbb{Z}))$ *si* $n/2$ *est pair, à* $O_h(H^n(X,\mathbb{Z})) \times \{\pm 1\}$ *si* $n/2$ *est impair.*

Soit u un difféomorphisme de X; comme $H^2(X,\mathbb{Z}) = \mathbb{Z}h$, on a $u^*h = \pm h$, d'où $u^*h^{n/2} = h^{n/2}$ ou $\pm h^{n/2}$ suivant que $n/2$ est pair ou impair. Observons qu'il existe effectivement des difféomorphismes u tels que $u^*h = -h$, par exemple la conjugaison complexe si X est définie sur $\mathbb{R}$. On est donc ramené à prouver que le groupe $O_h(H^n(X,\mathbb{Z}))$ est contenu dans $\mathrm{Im}(\pi)$. On sait déjà qu'il en est ainsi de $O_h^+(H^n(X,\mathbb{Z}))$, en vertu du Théorème 2; il suffit donc de construire un difféomorphisme dont l'image dans $O(H^n(X,\mathbb{Z}))$ appartient à O_h mais pas à O_h^+.

On peut supposer $d \geq 3$. Le réseau $H^n(X,\mathbb{Z})_o$ est alors complet (n° 2), donc contient un plan hyperbolique U. Par chirurgie [K-W], la décomposition $H^n(X,\mathbb{Z}) = U \oplus U^\perp$ est réalisée topologiquement par une décomposition en somme connexe $X = (S^n \times S^n) \# X'$. Soit s la symétrie de S^n par rapport à un équateur. Le difféomorphisme (s,s) de $S^n \times S^n$ préserve l'orientation et admet des points fixes; en le recollant avec l'identité de X', on obtient un difféomorphisme qui induit $-\mathrm{Id}$ sur U et Id sur $U^{\perp}$, donc préserve $h^{n/2}$ mais n'appartient pas à O_h^+, d'où la proposition.

PROPOSITION 5.- <u>Supposons</u> n <u>impair</u>. <u>Si</u> d <u>est impair et</u> $n \neq 1,3,7$, <u>l'image de</u> σ <u>est le groupe</u> $SpO(H^n(X,\mathbb{Z}),q_X)$. <u>Dans les autres cas, c'est le groupe</u> $Sp(H^n(X,\mathbb{Z}))$.

Si d est impair et $n \neq 1,3,7$, la forme q_X est invariante par difféomorphisme. Elle admet en effet la description suivante ([B], [W]) : toute classe $x \in H^n(X,\mathbb{Z})$ peut être représentée par une sphère S^n plongée dans X, et on a $q(x) = 0$ si et seulement si le fibré normal de S^n dans X est trivial. On a donc $\mathrm{Im}(\sigma) \subset SpO(H^n(X,\mathbb{Z}),q_X)$, d'où l'égalité compte tenu du Théorème 4.

Si d est pair, la proposition résulte du Théorème 2. Reste à traiter le cas $n = 1,3,7$. Il suffit de montrer que l'image de σ contient les transvections symplectiques s_δ, pour δ non divisible dans $H^n(X,\mathbb{Z})$. Soit ε un élément de $H^n(X,\mathbb{Z})$ tel que $\delta.\varepsilon = 1$, et soit $U = \mathbb{Z}\delta \oplus \mathbb{Z}\varepsilon$. Comme dans le cas n pair, la décomposition $H^n(X,\mathbb{Z}) = U \oplus U^{\perp}$ est réalisée topologiquement [W] par une décomposition $X = (S^n \times S^n) \# X'$. Soit v le difféomorphisme de $S^n \times S^n$ défini par $v(x,y) = (x,x.y)$, où le point désigne la multiplication des nombres complexes (resp. des quaternions, resp. des octonions) de norme 1. On vérifie facilement que le difféomorphisme obtenu en recollant v et l'identité de X' induit s_δ sur $H^n(X,\mathbb{Z})$, ce qui achève la démonstration de la proposition.

<u>Remarques</u>.- 1) Je suppose que les Propositions 4 et 5 peuvent aussi se déduire directement des décompositions en sommes connexes obtenues dans [K-W] et [W].

2) La Proposition 4 ne s'étend malheureusement pas au cas des surfaces, qui est certainement le plus intéressant pour les topologues. Il n'y a en effet aucune raison pour qu'un difféomorphisme de la surface X préserve la classe h. Pour $d = 3$, on a $\pi(\mathrm{Diff}^+(X)) = O(H^2(X,\mathbb{Z}))$ d'après un théorème de Wall [Wa]. Pour $d = 4$, on sait que l'image de π contient le sous-groupe $\mathrm{Ker}\,\sigma_{-1}$: cela résulte du fait que l'espace des modules des surfaces K3 marquées a deux composantes connexes, cf. [X], exposé XIII. Il semble que S. Donaldson sache prouver l'égalité $\pi(\mathrm{Diff}^+(X)) = \mathrm{Ker}\,\sigma_{-1}$ dans ce cas. J'ignore ce qui se passe pour $d \geq 5$.

5. <u>Intersections complètes</u>.

Les considérations qui précèdent s'étendent sans difficulté au cas des intersections complètes. Soient r un entier ≥ 2, et $\underline{d} = (d_1,\ldots,d_r) \in \mathbb{N}^r$, avec $d_i \geq 2$ pour tout i. Notons $U_{n,\underline{d}}$ l'ouvert de $\mathbb{P}^{N_1} \times \ldots \times \mathbb{P}^{N_r}$ (avec $N_i = \binom{n+r+d_i}{d_i}$) qui paramètre les intersections complètes lisses de r hypersurfaces de degrés $d_1,\ldots,d_r$ dans $\mathbb{P}^{n+r}$. Soient $u \in U_{n,\underline{d}}$, X l'intersection complète correspondante, $\Gamma_{n,\underline{d}}$ l'image de l'homorphisme de monodromie $\rho : \pi_1(U_{n,\underline{d}},u) \longrightarrow \mathrm{Aut}(H^n(X,\mathbb{Z}))$.

THEOREME 5.- Supposons n pair, et $\underline{d} \neq (2,2)$. Le groupe de monodromie $\Gamma_{n,\underline{d}}$ est le sous-groupe $O_h^+(H^n(X,\mathbb{Z}))$, d'indice 2 dans $O_h(H^n(X,\mathbb{Z}))$.

Pour $\underline{d} = (2,2)$ le groupe de monodromie est le groupe de Weyl $W(D_{2n+3})$, qui est égal à $O_h(H^n(X,\mathbb{Z}))$ (cf. [R]).

Reprenant la démonstration du Théorème 2, il suffit, pour chaque couple $(n,\underline{d})$, d'exhiber une intersection complète de dimension n et multidegré $\underline{d}$ admettant une singularité U_{12}. Si l'un des d_i, par exemple d_1, est ≥ 3, on prend la variété d'équations affines dans $\mathbb{C}^{n+r}$

$$x_1^3 + x_1^{d_1} + x_2^3 + \sum_{i=4}^{n+r} x_i^2 = 0 \quad ; \qquad x_i = x_4^2 + x_4^{d_i} \quad \text{pour } 4 \leq i \leq r+2.$$

Si $d_i = 2$ pour tout i (et $r \geq 3$), on prend pour équations

$$(x_1 + x_2)x_3 + \sum_{i=5}^{n+r} x_i^2 = 0 \quad ; \quad x_3 = x_1^2 - x_1x_2 + x_2^2 \quad ; \quad x_i = x_4^2 \quad \text{pour } 5 \leq i \leq r+2.$$

THEOREME 6.- Supposons $n = 2\nu - 1$, et $\underline{d} \neq (2,2)$. Soit p le nombre des degrés d_i qui sont pairs.

(i) Si $\binom{\nu+p-1}{\nu}$ est pair, il existe une forme quadratique canonique q_X mod. 2 sur $H^n(X,\mathbb{Z})$, et on a $\Gamma_{n,\underline{d}} = SpO(H^n(X,\mathbb{Z}), q_X)$.

(ii) Si $\binom{\nu+p-1}{\nu}$ est impair, le groupe de monodromie $\Gamma_{n,\underline{d}}$ est égal à $Sp(H^n(X,\mathbb{Z}))$.

Pour $\underline{d} = (2,2)$, la monodromie est celle de la famille universelle des courbes hyperelliptiques de genre ν [R] : le groupe $\Gamma_{n,\underline{d}}$ est formé des automorphismes symplectiques de $H^n(X,\mathbb{Z})$ dont la réduction mod. 2 appartient à un sous-groupe de $Sp(H^n(X,\mathbb{Z}/2))$ isomorphe au groupe symétrique $\mathfrak{S}_{n+3}$.

Démontrons le Théorème 6. On construit d'abord, pour tout $(n,\underline{d})$, une intersection complète de dimension n et multidegré $\underline{d}$ admettant une singularité E_6. Si l'un des d_i, disons d_1, est ≥ 3, on prend pour équations affines

$$x_1^3 + x_1^{d_1} + \sum_{i=3}^{n+r} x_i^2 = 0 \quad ; \qquad x_i = x_2^2 + x_2^{d_i} \quad \text{pour } 3 \leq i \leq r+1.$$

Si $d_i = 2$ pour tout i (et $r \geq 3$), on prend

$$x_1x_2 + \sum_{i=4}^{n+r} x_i^2 = 0 \quad ; \quad x_2 = x_1^2 \quad ; \quad x_i = x_3^2 \quad \text{pour } 4 \leq i \leq r+1.$$

On est donc dans l'une des situations (i) ou (ii) du Théorème 3. Si $\binom{\nu+p-1}{\nu}$ est pair, il existe une forme quadratique mod. 2 q_X sur $H^n(X,\mathbb{Z})$, invariante par déformation [B] : cela entraîne (i). Il reste à éliminer, lorsque $\binom{\nu+p-1}{\nu}$

est impair, l'existence d'une telle forme; il suffit pour cela, comme dans la démonstration du Théorème 4, de trouver pour chaque $(n,\underline{d})$ une situation $Z \subset X_o$, où X_o est une intersection complète de dimension n et multidegré $\underline{d}$ avec des points doubles ordinaires, et Z une sous-variété lisse de dimension ν passant par un nombre impair de ces points doubles.

Soit (A_{ij}) une matrice de formes homogènes sur $\mathbb{P}^{\nu}$ à r lignes et $\nu+r-1$ colonnes, avec $\deg A_{ij} = d_i - 1$. Prenons pour X_o la sous-variété de $\mathbb{P}^{n+r}$ définie par les équations

$$\sum_{j=1}^{\nu+r-1} T_{\nu+j} A_{ij}(T_o,\dots,T_\nu) = 0 \qquad (1 \leq i \leq r),$$

et pour Z le sous-espace linéaire $T_{\nu+1} = \dots = T_{n+r} = 0$ de $\mathbb{P}^{n+r}$. Lorsque les formes A_{ij} sont assez générales, on déduit du critère jacobien que les seuls points singuliers de X_o sont des points doubles ordinaires, situés sur Z et définis par l'équation

$$\operatorname{rg}(A_{ij}) \leq r-1.$$

Le nombre N des points de $Z(=\mathbb{P}^{\nu})$ vérifiant cette équation est donnée par une formule classique de géométrie énumérative, exprimée en langage moderne par Porteous [P] :

$$N = \det((c_{j-i+1})_{1\leq i,j\leq \nu}) ,$$

où c_q désigne la q-ième classe de Chern du fibré $\sum_{i=1}^{r} \mathcal{O}_{\mathbb{P}^\nu}(d_i-1)$.

Dans $H^*(\mathbb{P}^{\nu}, \mathbb{Z}/2)$, on a

$$c(\sum_i \mathcal{O}_{\mathbb{P}^\nu}(d_i-1)) = c(\mathcal{O}_{\mathbb{P}^\nu}(1)^p) = (1+c_1(\mathcal{O}_{\mathbb{P}}(1)))^p,$$

d'où la congruence

$$N \equiv \det(({\textstyle\binom{p}{j-i+1}})_{1\leq i,j\leq \nu}) \pmod 2.$$

On conclut la démonstration à l'aide du lemme élémentaire suivant :

Lemme.- On a $\det(({\textstyle\binom{p}{j-i+1}})_{1\leq i,j\leq \nu}) = \binom{p+\nu-1}{\nu}$.

Fixons p, et notons Δ_ν le déterminant à calculer $(\nu \geq 1)$. En développant Δ_ν suivant les colonnes, on obtient la relation

$$\Delta_\nu - \binom{p}{1}\Delta_{\nu-1} + \binom{p}{2}\Delta_{\nu-2} + \dots + (-1)^{\nu}\binom{p}{\nu} = 0.$$

Raisonnant par récurrence sur ν, il suffit de vérifier que cette relation est satisfaite lorsqu'on remplace Δ_i par $\binom{p+i-1}{i} = (-1)^i\binom{-p}{i}$. Mais l'expression

$$\binom{-p}{\nu} + \binom{-p}{\nu-1}\binom{p}{1} + \ldots + \binom{-p}{1}\binom{p}{\nu-1} + \binom{p}{\nu}$$

n'est autre que la coefficient de T^{ν} dans le développement en série de $(1+T)^{-p}(1+T)^{p}$, d'où le lemme et le théorème.

Les Propositions 1 à 5 du n° 4 se généralisent immédiatement au cas des intersections complètes; je laisse les détails au lecteur.

BIBLIOGRAPHIE

[A] V.I. ARNOL'D : Remarks on the stationary phase methods and Coxeter numbers. Russian Math. Surveys 28, 5 (1973), 19-48.

[B] W. BROWDER : Complete intersections and the Kervaire invariant. Algebraic topology, Aarhus 1978. Springer Lecture Notes 763 (1979), 88-108.

[D] P. DU VAL : On isolated singularities which do not affect the conditions of adjunction, III. Proc. Cambridge Phil. Soc. 30 (1934), 483-491.

[E] W. EBELING : An arithmetic characterisation of the symmetric monodromy groups of singularities. Inventiones math. 77 (1984), 85-99.

[J] W.A.M. JANSSEN : Skew-symmetric vanishing lattices and their monodromy group. Math. Ann. 266 (1983), 115-133; 272 (1985), 17-22.

[K-W] R. KULKARNI and J. WOOD : Topology of non-singular complex hypersurfaces. Adv. in Math. 35 (1980), 239-263.

[P] I.R. PORTEOUS : Simple singularities of maps. Proc. of Liverpool Singularities Symposium, Springer Lecture Notes 192 (1971), 286-307.

[R] M. REID : The complete intersection of two or more quadrics. Thesis, Cambridge (1972).

[Wa] C.T.C. WALL : Diffeomorphisms of 4-manifolds. J. London Math. Soc. 39 (1964), 131-140.

[W] J. WOOD : Removing handles from non-singular algebraic hypersurfaces in $\mathbb{C}\mathbb{P}_{n+1}$. Inventiones Math. 31 (1976), 1-6.

[X] Séminaire Palaiseau : Géométrie des surfaces K3, modules et périodes. Astérisque 126 (1985).

Mathématiques, Bât. 425, Université Paris-Sud
F-91405 ORSAY CEDEX

COMPLETE FAMILIES OF STABLE VECTOR BUNDLES OVER $\mathbb{P}_2$

A. HIRSCHOWITZ
Laboratoire de Mathématiques
U.A. CNRS n°168
Parc Valrose
F-06034 NICE CEDEX

K. HULEK
Mathematisches Institut
Universität Bayreuth
Postfach 3008
D-8580 BAYREUTH

O. INTRODUCTION

Let $\overline{M}(c_1,c_2)$ be the moduli scheme of semi-stable rank 2 sheaves over $\mathbb{P}_2$ with Chern classes c_1 and c_2 and let $M(c_1,c_2) \subseteq \overline{M}(c_1,c_2)$ be the open set of stable locally free sheaves, i.e. of stable rank 2 vector bundles. $\overline{M}(c_1,c_2)$ is an irreducible projective variety of dimension $4c_2-c_1^2-3$. The Zariski open set $M(c_1,c_2)$ is smooth and if $c_1^2-4c_2 \not\equiv 0 \mod 8$ this is also true for $\overline{M}(c_1,c_2)$. In this case the boundary $S := \overline{M}(c_1,c_2) \setminus M(c_1,c_2)$ is an irreducible divisor in $\overline{M}(c_1,c_2)$. (For these statements see Maruyama [8] and Strømme [11]).

The problem which we should like to propose is the following :

Question 1 : What is the maximal dimension of a complete subvariety $X \subseteq M(c_1,c_2)$?

This is equivalent to

Question 2 : What is the maximal dimension of a complete non-degenerate family of stable rank 2 vector bundles over $\mathbb{P}_2$?

By a complete non-degenerate family we mean a vector bundle E over $\mathbb{P}_2 \times X$ where X is an irreducible complete variety, such that for every $x \in X$ the bundle $E_x = E|\,\mathbb{P}_2 \times \{x\}$ is stable and the induced map $X \to M(c_1,c_2)$; $x \mapsto E_x$ has generically finite fibres. The equivalence of questions 1 and 2 follows from the existence of a universal $\mathbb{P}_1$-bundle over $M(c_1,c_2)$. If $c_1^2 - 4c_2 \not\equiv 0 \mod 8$ this $\mathbb{P}_1$-bundle comes from a vector bundle (Le Potier [7]). If $c_1^2 - 4c_2 \equiv 0 \mod 8$ one can take a suitable double cover of $M(c_1,c_2)$ such that the pullback of the universal $\mathbb{P}_1$-bundle comes from a vector bundle.

There are two reasons why we are interested in this problem.

1) An answer to the above questions provides in some sense a "measure" in how far the moduli scheme $M(c_1,c_2)$ is from either being affine or projective.

2) If E is a stable rank 2 bundle over $\mathbb{P}_n$, $n \geq 3$ with $c_1(E)=0$ or -1 and $c_2(E)=c_2$ then the restriction of E to a general plane H is known to be (semi-)stable (cf. Barth [2]). Hence an upper bound on the

dimension of complete subvarieties in a suitable quotient $M(c_1,c_2)/PGL(3)$ would imply a lower bound on the dimension of the variety H of planes such that the restriction $E|H$ is unstable (with respect to the PGL(3)-action or in the usual sense). Although there is not necessarily a direct connection between our question 1 and this problem, they certainly have the same flavour.
We should like to mention that for a so-called general instanton bundle E over $\mathbb{P}_3$ with $c_2 \geq 5$ the restriction to every plane H in $\mathbb{P}_3$ is stable (cf. Brun-Hirschowitz [5]).
This restriction is probably also stable with respect to the PGL(3)-action, hence leading to complete 3-dimensional families in the quotient.

We are very far from being able to answer the above questions. Nevertheless we feel that this problem should be posed. The main point of this note is to give a first result in this direction. We shall prove

Theorem : **If $c_1=0$ and $c_2 \geq 3$ or $c_1=-1$ and $c_2 \geq 2$ then complete rational curves $R \subsetneq M(c_1,c_2)$ exist.**

Our methods are very simple and explicit. Unfortunately, however, it seems to be difficult to extend them to obtain complete families of dimension greater than one.

The contents of this paper are as follows :
In section 1 we shall construct the curves R explicitly.
In section 2 we show that the curves constructed in section 1 can be deformed in $M(c_1,c_2)$ in such a way as to go through a general point of the moduli scheme. In section 3 we shall relate our construction to the results of Strømme [11] and compute the numerical class of our curves in $M(c_1,c_2)$ in case $c_1^2-4c_2 \not\equiv 0 \bmod 8$. In section 4 we shall discuss some special cases.
Throughout this paper we shall work with Gieseker stability. A coherent rank 2 sheaf E is called (semi-)stable if it is torsion-free and if for every rank 1 subsheaf $E' \subseteq E$ the following inequality holds for sufficiently large integers k :

$$2\chi(E'(k)) \underset{(=)}{<} \chi(E(k)).$$

If $c_1^2 - 4c_2 \not\equiv 0 \bmod 8$ then every semi-stable sheaf is stable.

For rank 2 bundles over $\mathbb{P}_n$ Gieseker stability and Mumford-Takemoto stability coincide.

Throughout we shall work over the complex field $\mathbb{C}$ but all our arguments seem to extend to the case of any algebraically closed field k.

* *We would like to thank A. Strømme for an interesting discussion concerning section 3 of this paper.* *

1. THE BASIC CONTRUCTION

Let $Y=\{y_1,\ldots,y_k\}\subseteq \mathbb{P}_2$ be a set of k pointwise different points. We first want to recall the basic relation between such sets Y and rank 2 vector bundles which is given by

1.1 Proposition : Let $\ell=1$ or 2. Then every isomorphism $\det N_{Y/\mathbb{P}_2}\cong \mathcal{O}_{\mathbb{P}_2}(\ell)|Y$ defines an extension.

$$\textbf{(1)}\qquad 0 \longrightarrow \mathcal{O}_{\mathbb{P}_2} \longrightarrow E(1) \longrightarrow \mathcal{I}_Y(\ell) \longrightarrow 0.$$

Here E is a rank 2 bundle with Chern classes $c_1(E)=\ell-2$ and $c_2(E)=k-\ell+1$.

Proof : can be founded e.g. in [10] p.90ff.

If E arises from the extension (1) then according to Barth [2] p.132 it is stable if and only if

$$h^{o}(E) = h^{o}(\mathcal{I}_Y(\ell-1)) = 0.$$

This is always the case if $\ell=1$. If $\ell=2$ this is fulfilled if and only if $n\geq 3$ and the points $x_1,\ldots,x_n$ are not colinear.

Next we consider the threefold

$$X: = \mathbb{P}_2\times\mathbb{P}_1$$

together with its natural projections

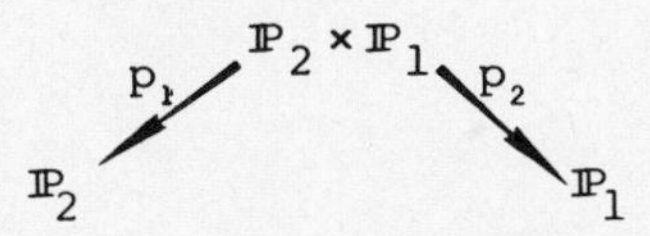

We shall think of X as the trivial $\mathbb{P}_2$-bundle over $\mathbb{P}_1$. Clearly

$$\operatorname{Pic} X = p_1^*\operatorname{Pic}\mathbb{P}_2 \oplus p_2^*\operatorname{Pic}\mathbb{P}_1 = \mathbb{Z}\oplus\mathbb{Z}$$

and we shall employ the notation

$$\mathcal{O}_X(a,b):=p_1^*\mathcal{O}_{\mathbb{P}_2}(a)\otimes p_2^*\mathcal{O}_{\mathbb{P}_1}(b).$$

Before we can prove the main result of this section we need some lemmas. this is the first of them :

1.2 Lemma : Let $M = \mathcal{O}_X(a,b)$ with $a+b=3$. Then

(i) $h^1(M^{-1})=0$ if and only if $a\geq 1$ (and hence $b\leq 2$).

(ii) $h^2(M^{-1})=0$ if and only if $a\leq 2$ (and hence $b\geq 1$).

Proof : (i) By the Künneth formula we have

$$\begin{aligned} h^1(M^{-1}) &= h^1(\mathcal{O}_X(-a_1-b)) \\ &= h^1(\mathcal{O}_{\mathbb{P}_2}(-a))\cdot h^{o}(\mathcal{O}_{\mathbb{P}_1}(-b)) + h^{o}(\mathcal{O}_{\mathbb{P}_2}(-a))\cdot h^1(\mathcal{O}_{\mathbb{P}_1}(-b)) \\ &= h^{o}(\mathcal{O}_{\mathbb{P}_2}(-a))\cdot h^1(\mathcal{O}_{\mathbb{P}_1}(-b)). \end{aligned}$$

From this the assertion follows immediately.

(ii) This follows in a similar way from another application of the Künneth formula.

Now let $\varphi: \mathbb{P}_1 \hookrightarrow \mathbb{P}_2$ be a linear embedding of $\mathbb{P}_1$ as a line in $\mathbb{P}_2$. Then φ defines a section of X in the following way :

$$\bar{\varphi} : \mathbb{P}_1 \longrightarrow X = \mathbb{P}_2 \times \mathbb{P}_1$$
$$t \longmapsto (\varphi(t), t) .$$

the image

$$L := \bar{\varphi}(\mathbb{P}_1)$$

is a smooth curve in X. Let $N_{L/X}$ be its normal bundle.

1.3 **Lemma :** $$N_{L/X} = \mathcal{O}_{\mathbb{P}_1}(1) \oplus \mathcal{O}_{\mathbb{P}_1}(2).$$

Proof : We shall first show that the degree of $N_{L/X}$ is 3. From the exact sequence

$$0 \longrightarrow T_L \longrightarrow T_X|L \longrightarrow N_{L/X} \longrightarrow 0$$

it follows that

$$\det N_{L/X} = T_L^{-1} \otimes \det T_X|L.$$

Since

$$(\det T_X)^{-1} = \omega_X = p_1^* \omega_{\mathbb{P}_2} \otimes p_2^* \omega_{\mathbb{P}_1} = \mathcal{O}_X(-3,-2)$$

we get

$$\det N_{L/X} = \mathcal{O}_{\mathbb{P}_1}(-2) \otimes \mathcal{O}_X(3,2)|L = \mathcal{O}_{\mathbb{P}_1}(3).$$

If we denote the image of the embedding $\mathbb{P}_1 \hookrightarrow \mathbb{P}_2$ by $\mathbb{P}_1$ too, we have

$$L \subseteq \mathbb{P}_1 \times \mathbb{P}_1 \subseteq X .$$

Here L is the diagonal of the quadric $\mathbb{P}_1 \times \mathbb{P}_1$. Hence we get an exact sequence

$$\begin{array}{ccccccc} 0 \longrightarrow & N_{L/\mathbb{P}_1\times\mathbb{P}_1} & \longrightarrow & N_{L/X} & \longrightarrow & N_{\mathbb{P}_1\times\mathbb{P}_1/X}|L & \longrightarrow 0 . \\ & \| & & & & \| & \\ & \mathcal{O}_{\mathbb{P}_1}(2) & & & & \mathcal{O}_{\mathbb{P}_1}(1) & \end{array}$$

Since this sequence must split we are done.

For what follows we have to consider k-tuples of linear embeddings $\varphi_i: \mathbb{P}_1 \hookrightarrow \mathbb{P}_2$; $i=1,\dots,k$ and their associated sections $L_i := \bar{\varphi}_i(\mathbb{P}_1) \subseteq X$. The linear embeddings $\mathbb{P}_1 \hookrightarrow \mathbb{P}_2$ are parametrized by the open set $U \subseteq \mathbb{P}_5 = \mathbb{P}(\mathrm{Hom}(\mathbb{C}^2, \mathbb{C}^3))$ which consist of linear mappings of rank 2. Hence a k-tuple of sections $L_1,\dots,L_k$ is given by a point in U^k (resp. S^kU) and it makes sense to talk about "almost all" k-tuples of sections (i;e. all k-tuples outside some Zariski-closed set in U^k).

1.4 **Lemma : (i)** For almost all k-tuples of linear embeddings $\varphi_1,\dots,\varphi_k: \mathbb{P}_1 \hookrightarrow \mathbb{P}_2$ the corresponding sections $L_1,\dots,L_k$ are disjoint. In particular, the union $Y := \bigcup_{i=1}^{k} L_i \subseteq X$ is a smooth 2-codimensional subvariety.

(ii) Assume $k \geq 4$. Then almost all k-tuples of embeddings

$\varphi_1,\ldots,\varphi_k: \mathbb{P}_1 \hookrightarrow \mathbb{P}_2$ have the property that for each $t \in \mathbb{P}_1$ the points $\varphi_1(t),\ldots,\varphi_k(t) \in \mathbb{P}_2$ are not colinear.

Proof : (i) We shall give a proof by induction. The assertion is trivial for **k**=1. Assume that $\varphi_1,\ldots,\varphi_k$ are such that $L_1,\ldots,L_k$ are disjoint. Now look at embeddings $\varphi_{k+1}: \mathbb{P}_1 \hookrightarrow \mathbb{P}_2$. We can assume that $\varphi_{k+1}(\mathbb{P}_1) \neq \varphi_i(\mathbb{P}_1)$ for $i=1,\ldots,k$. Hence we have

$$\varphi_{k+1}(\mathbb{P}_1) \cap \varphi_i(\mathbb{P}_1) = P_i$$

and there are uniquely determined points $t_1,\ldots,t_k \in \mathbb{P}_1$ and $t^1,\ldots,t^k \in \mathbb{P}_1$ such that

$$\varphi_i(t_i) = \varphi_{k+1}(t^i) = P_i.$$

The condition that L_{k+1} is disjoint from $\bigcup_{i=1}^{k} L_i$ is now equivalent so

$$t_i \neq t^i.$$

This is clearly an open non-empty condition.

(ii) It is enough to prove this for k=4. To begin with , choose $\varphi_1,\ldots,\varphi_3: \mathbb{P}_1 \hookrightarrow \mathbb{P}_2$ in such a way that $L_1,\ldots,L_3$ are disjoint and that the points $\varphi_1(t_o),\varphi_2(t_o),\varphi_3(t_o)$ for some $t_o \in \mathbb{P}_1$ are not colinear. The set of points t such that $\varphi_1(t),\ldots,\varphi_3(t)$ are colinear is then given by

$$\det(\varphi_1(t),\varphi_2(t),\varphi_3(t)) = 0.$$

Since the φ_i are linear and since this determinent is not identically zero, there are (properly counted) three such points, say t_1, t_2 and t_3. The points $\varphi_i(t_j)$ for fixed j and $i=1,\ldots,3$ determine a line L_j; $j=1,\ldots,3$. It is now sufficient to choose φ_4 in such a way that

$$\varphi_4(t_j) \notin L_j \ ; \ j=1,\ldots,3.$$

This concludes the proof.

We are now ready to state the main result of this section.

1.5 <u>Theorem</u> : **Assume $c_1=0$ and $c_2 \geq 3$ or $c_1=-1$ and $c_2 \geq 2$. Then there exist complete, non-degenerate, 1-dimensional families of stable rank 2 vector bundles over $\mathbb{P}_2$ with these Chern classes. More precisely there exist rank 2 vector bundles over $\mathbb{P}_2 \times \mathbb{P}_1$ such that for each $t \in \mathbb{P}_1$ the bundle $E_t := E|\mathbb{P}_2 \times \{t\}$ is stable with $c_1(E_t)=c_1$ and $c_2(E_t)=c_2$ and such that the map $\mathbb{P}_1 \longrightarrow M(c_1,c_2)$, $t \longmapsto E_t$ is non-constant. In particular there exist complete rational curves $R \subseteq M(c_1,c_2)$.**

Proof : (i) We shall first prove the theorem for $c_1=-1$. By lemma (1.4) we can choose for each $k \geq 2$ linear embeddings $\varphi_1,\ldots,\varphi_k: \mathbb{P}_1 \rightarrow \mathbb{P}_2$ such that $Y = \bigcup_{i=1}^{k} L_i$ is smooth. By lemma (1.3) we can moreover choose an isomorphism

$$\det N_{Y/X} \cong \mathcal{O}_X(1,2)|Y.$$

Since $h^2(\mathcal{O}_X(-1,-2)) = h^1(\mathcal{O}_X(-1,-2)) = 0$ by lemma (1.2) this determines -via the Serre construction (cf.[10]p.90ff)- a unique extension

$$\text{(2)} \qquad 0 \longrightarrow \mathcal{O}_X \longrightarrow E(1,0) \longrightarrow J_Y(1,2) \longrightarrow 0 .$$

For each $t \in \mathbb{P}_1$ this restricts to an extension

$$\text{(3)} \qquad 0 \longrightarrow \mathcal{O}_{\mathbb{P}_2} \longrightarrow E(1) \longrightarrow J_{Y_t}(1) \longrightarrow 0$$

where $Y_t = Y \cap (\mathbb{P}_2 \times \{t\})$ consists of k different points. Hence E_t is stable for all $t \in \mathbb{P}_1$ and we have

$$c_1(E_t) = -1, \quad c_2(E_t) = k .$$

It remains to show that we can arrange Y in such a way that the induced morphism $\mathbb{P}_1 \longrightarrow M(-1,c_2)$ is not constant. To see this, notice the following. If $Y_t = \{y_1^t,\dots,y_k^t\}$ then the line L_{ij}^t joining the points y_i^t and y_j^t is a jumping line of E_t, i.e..

$$E_t|L_{ij}^t = \mathcal{O}_{\mathbb{P}_1}(-k-1) \oplus \mathcal{O}_{\mathbb{P}_1}(k) \quad \text{with } k \geq 1 .$$

On the other hand one can deduce from the extension (3) that the general line is not a jumping line of E_t. Hence if we choose $\varphi_1,\dots,\varphi_k$ general enough we can always find points t_o and t_1 such that the bundles E_{t_0} and E_{t_1} have different jumping lines and are, therefore, not isomorphic.

(ii) The case $c_1=0$ can be treated similarly. If $k \geq 4$ we can choose embeddings $\varphi_1,\dots,\varphi_k : \mathbb{P}_1 \longrightarrow \mathbb{P}_2$ such that the conditions **(i)** and **(ii)** of lemma 1.4 are fulfilled. As before this gives rise to an extension

$$0 \longrightarrow \mathcal{O}_X \longrightarrow E(1,0) \longrightarrow \mathcal{J}_Y(2,1) \longrightarrow 0 .$$

Since the points $\varphi_1(t),\dots,\varphi_k(t)$ are never colinear the bundles E_t are always stable. Their Chern classes are

$$c_1(E_t)=0 , \quad c_2(E_t) = k-1 .$$

By the same argument as before we can arrange $L_1,\dots,L_k$ in such a way that the induced map $\mathbb{P}_1 \longrightarrow M(0,c_2)$ is not constant.

<u>Remark</u> : We shall see in section 4 that there exist no such families of bundles if $c_1=-1$, $c_2=1$ or $c_1=0$, $c_2=2$.

2. CURVES THROUGH A GENERAL POINT OF THE MODULI SCHEME

The families of bundles which we have constructed so far have a very special property. To explain this recall that for every rank 2 bundle E with Chern classes c_1 and c_2 the <u>normalized bundle</u> E_{norm} is defined by

$$E_{norm} := E(-[\tfrac{c_1+1}{2}]) .$$

Here $[x]$ denotes as usual the greatest integer less than or equal to x. E_{norm} has the property that $c_1(E_{norm})=-1$ or $c_1(E_{norm})=0$ depending on whether c_1 is odd or even.

We make the following

2.1 **Definition :** $Z(c_1,c_2):=\{E \in M(c_1,c_2)\ ;\ h^o(E_{norm}(1))\neq 0\}$.
In other words if $c_1=0$ or -1 then $Z(c_1,c_2)$ is the Zariski closed set of all bundles whose first twist has non-zero sections. J.Brun [4] p.469 showed that

$$Z(c_1,c_2) \neq M(c_1,c_2) \Leftrightarrow \tfrac{1}{2}c_1^{norm}(c_1^{norm}+5) - c_2^{norm}+6 \leq 0 .$$

Here (c_1^{norm}, c_2^{norm}) are the Chern classes of a normalized bundle E_{norm} where $E \in M(c_1,c_2)$. This implies that if $c_1=0$, $c_2\geq 6$ or $c_1=-1$, $c_2\geq 4$ then $Z(c_1,c_2) \neq M(c_1,c_2)$.

Now let E be a rank 2 vector bundle over $\mathbb{P}_2\times\mathbb{P}_1$ as constructed in the previous section. Then clearly for every point $t \in \mathbb{P}_1$ the bundle $E_t \in Z(c_1,c_2)$. Hence the induced curves R in the moduli scheme $M(c_1,c_2)$ lie entirely in $Z(c_1,c_2)$. The above construction does not generalize readily to curves which are not contained in $Z(c_1,c_2)$. Nevertheless such curves exist. In fact one can find curves through a general point of $M(c_1,c_2)$, i.e. curves which pass through a point which belongs to a vector bundle with natural cohomology. Here E is said to have natural cohomology if for each $k\in\mathbb{Z}$ at most one of the cohomology groups $h^i(E(k))$ is non-zero (for details see Brun [4]).

The above assertion follows from the following

2.2 **Proposition :** (i) $H^2(\mathrm{End}(E))=0$, i.e. the base space of the versal deformation is smooth.
(ii) For every point $t \in \mathbb{P}_1$ the restriction map $H^1(\mathrm{End}\ E)\longrightarrow H^1(\mathrm{End}\ E_t)$ is surjective.

Proof : We set $F:=E(1,0)$. Then $\mathrm{End}\ F\cong \mathrm{End}\ E$ and we have the following exact sequences :

$$\textbf{(6)}\quad 0 \longrightarrow \mathcal{O}_X \longrightarrow F \longrightarrow \mathcal{I}_Y(a,b) \longrightarrow 0$$

$$\textbf{(7)}\quad 0 \longrightarrow F^* \longrightarrow \mathrm{End}\ F \longrightarrow F^* \otimes \mathcal{I}_Y(a,b) \longrightarrow 0$$

$$\textbf{(8)}\quad 0 \longrightarrow \mathcal{I}_Y \longrightarrow \mathcal{O}_X \longrightarrow \mathcal{O}_Y= \bigoplus_{i=1}^{k} \mathcal{O}_{\mathbb{P}} \longrightarrow 0 .$$

Here (a,b) is $(1,2)$ or $(2,1)$. Moreover we have an isomorphism

$$F \cong F^*(a,b) .$$

<u>Step</u> **1 :** $h^2(F^*)=0$. Since $F^*\cong F(-a,-b)$ and because of sequence (6) it is enough to show that $h^2(\mathcal{O}_X(-a,-b))=0$ and $h^2(\mathcal{I}_Y)=0$. The first statement follows from lemma 1.2, the second statement follows

from (8) since $h^1(\mathcal{O}_{\mathbb{P}})=0$ and $h^2(\mathcal{O}_X)=0$. The latter is a consequence of the Künneth formula.

Step **2** : $h^2(F^* \otimes \mathcal{I}_Y(a,b))=0$. Tensoring (8) with $F^*(a,b)$ we get

$$0 \longrightarrow F^* \otimes \mathcal{I}_Y(a,b) \longrightarrow F^*(a,b) \longrightarrow F^*(a,b)|Y \longrightarrow 0 .$$

It follows from sequence (6) that

$$F|Y \cong N^*_{Y/X}(a,b) .$$

Hence by lemma 1.3 we get

$$F^*(a,b)|Y \cong N_{Y/X} = \bigoplus_{i=1}^{n} (\mathcal{O}_{\mathbb{P}_1}(1) \oplus \mathcal{O}_{\mathbb{P}_1}(2)) .$$

This shows that $h^1(F^*(a,b)|Y) = 0$.

Next we claim that $h^2(F^*(a,b)) = h^2(F) = 0$. Using (6) and the fact that $h^2(\mathcal{O}_X)=0$ this will follow from $h^2(\mathcal{I}_Y(a,b))=0$. This, however, can be deduced from (8) since $h^1(\mathcal{O}_Y(a,b)) = \sum_{i=1}^{k} h^1(\mathcal{O}_{\mathbb{P}_1}(3))=0$ and $h^2(\mathcal{O}_X(a,b))=0$ by the Künneth formula

Steps 1 and 2 give a proof for (i). To prove (ii) we look at the exact sequence

$$0 \longrightarrow (\text{End } E)(0,-1) \longrightarrow \text{End } E \longrightarrow \text{End } E_t \longrightarrow 0 .$$

It will be sufficient to show that

$$h^2((\text{End } E)(0,-1)) = h^2((\text{End } F)(0,-1)) = 0.$$

Since this can be proved by an argument very similar to the one just given for (i), we shall omit the details.

From the above proposition we get immediately

2.3 **Corollary :** If $c_1=0$, $c_2 \geq 6$ or $c_1=-1$, $c_2 \geq 4$ then there exist complete rational curves R in $M(c_1,c_2)$ which are not contained in $Z(c_1,c_2)$.

Proof : In view of 2.2(i) it follows from 2.2(ii) that the versal deformation of E induces the versal deformation of E_t. On the other hand, by openness of stability a small deformation of E still defines a complete curve in $M(c_1,c_2)$.

Remark : It is not hard to check that the bundle E on $X = \mathbb{P}_2 \times \mathbb{P}_1$ is Mumford-Takemoto stable with respect to the ample line bundle $\mathcal{O}_X(1,1)$. Moreover

$$h^o(\text{End } E) = 1 \ , \ h^3(\text{End } E) = 0.$$

One can, therefore, use Riemann-Roch to compute

$$h^1(\text{End } E) = 1-\chi(\text{End } E) = \begin{cases} 10c_2-3 & \text{if } c_1=0 \\ 10c_2-10 & \text{if } c_1=-1 . \end{cases}$$

3. INTERSECTING DIVISORS IN $\overline{M}(c_1,c_2)$

If $c_1^2 - 4c_2 \not\equiv 0 \mod 8$ the projective variety $\overline{M}(c_1,c_2)$ is smooth. In [11] Strømme computed its Picard group. He also showed that the boundary $S=\overline{M}(c_1,c_2)\setminus M(c_1,c_2)$ is an irreducible divisor and computed its class in Pic $\overline{M}(c_1,c_2)$. Since the curves R lie entirely in $M(c_1,c_2)$ the intersection number S.R must be zero.

Here we shall more generally compute the intersection number D.R with any class D in Pic $\overline{M}(c_1,c_2)$. In other words we compute the numerical class of R. This gives also some information in the case $c_1^2 - 4c_2 \equiv 0 \mod 8$. Our ideas and notation follow closely [11].

In order to recall Strømme's results we start with

3.1 **Definition :** Let E be a normalized rank 2 sheaf on $\mathbb{P}_2$. A line L (resp.a conic C) $\subseteq \mathbb{P}_2$ is called a jumping line (resp. jumping conic) if $h^1(E(-c_1-1)|L) \neq 0$ (resp. $h^1(E|C) \neq 0$).

Remarks : (i) The definition of jumping line is the standard one. On the other hand one should point out that if $c_1 = 0$ a smooth conic $\mathbb{P}_1 \cong C \subseteq \mathbb{P}_2$ is a jumping conic according to this definition if and only if

$$E|C = \mathcal{O}_{\mathbb{P}_1}(-k) \oplus \mathcal{O}_{\mathbb{P}_1}(k) \quad \text{with } k \geq 2 .$$

(ii) If E is semi-stable the Grauert-Mülich theorem implies the general line (resp.conic) is not a jumping line (conic).

Strømme proved the following :

3.2 Theorem (Strømme) : **Assume that** $c_1=-1$ **(resp.** $c_1=0$) **and that** $c_2 = n \geq 2$ **(resp.** $c_2 = n \geq 3$ **is odd). Then**

(i) Pic $\overline{M}(c_1,c_2)$ **is freely generated by two generators** ε **and** δ **(resp.** φ **and** ψ**).**

(ii) **Consider the following subsets of** $\overline{M}(c_1,c_2)$ **:**

D_1 := **{sheaves with a given jumping conic (line)}.**

D_2 := **{sheaves with a jumping line (conic) passing through 1(resp.3) given points}.**

Then D_1 **is the support of a reduced divisor in the linear system** $|\varepsilon|$ **(resp.** $|\varphi|$**) and** D_2 **is the support of a reduced divisor in the linear system** $|\delta|$ **(resp.** $|\tfrac{1}{2}(n-1)\psi|$**).**

(iii) The boundary $S=\overline{M}(c_1,c_2)\setminus M(c_1,c_2)$ **is the support of a reduced and irreducible divisor whose class is** $n\varepsilon - 2\delta$ **(resp.**$(2n-1)\varphi - \psi$**).**

Proof : See [11].

For our application we have to describe the divisors D_1 and D_2 more closely. Let X be any smooth projective variety and let E be a rank 2 bundle over $\mathbb{P}_2 \times X$ such that $E_t := E|\mathbb{P}_2 \times \{t\}$ is stable for all $t \in X$ and $c_1(E_t) = -1$ or 0. We set $n := c_2 := c_2(E_t)$. We have the natural projections

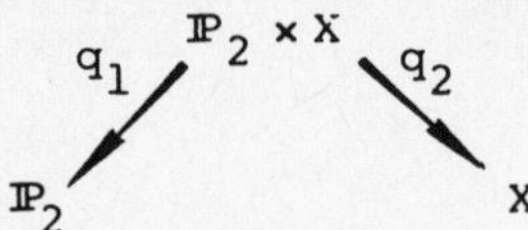

and set

$$\tau := q_1^* \mathcal{O}_{\mathbb{P}_2}(1) .$$

Since E_t is stable it follows that $h^i(E_t(-\ell\tau)) = 0$ for $i \neq 1$ and $0 \leq \ell \leq 2$ (see e.g. [11] p.407). Hence the direct image sheaves $R^1 q_{2*} E(-\ell\tau)$ are locally free of rank $h^1(E_t(-\ell\tau)) = -\chi(E_t(-\ell\tau))$ for $0 \leq \ell \leq 2$. Following [11] we set

$$\mathcal{A} := R^1 q_{2*} E(-2\tau) , \quad \mathcal{B} := R^1 q_{2*} E(-\tau), \quad \mathcal{C} := R^1 q_{2*} E .$$

A simple application of Riemann-Roch shows

$$\operatorname{rank} \mathcal{A} = n + c_1, \quad \operatorname{rank} \mathcal{B} = n, \quad \operatorname{rank} \mathcal{C} = n - 2 - c_1 .$$

It is now easy to describe D_1. Note that a conic C (a line L) is a jumping conic (line) of E_t if and only if the map

$$\alpha(t,C): H^1(E_t(-2)) \longrightarrow H^1(E_t)$$
$$(\text{resp. } \alpha(t,L): H^1(E_t(-2)) \longrightarrow H^1(E_t(-1)))$$

which is given by multiplication with the equation of C (resp. L) fails to be an isomorphism. Varying t we get morphisms

$$\alpha(C): \mathcal{A} \longrightarrow \mathcal{C} \ (\text{resp. } \alpha(L): \mathcal{A} \longrightarrow \mathcal{B}) .$$

3.3 **Lemma** : The class of D_1 in Pic X is given by

$$[D_1] = c_1(\mathcal{C}) - c_1(\mathcal{A}) \quad (\text{resp. } [D_1] = c_1(\mathcal{B}) - c_1(\mathcal{A})).$$

Proof : This is clear since $D_1 = \{\det \alpha(C) = 0\}$ (resp. $D_1 = \{\det \alpha(L) = 0\}$).

To describe the divisor D_2 is more difficult. To do this let Q be the projective line (plane) of lines (conics) through 1 (resp. 3) given points. In general (see [11]) the variety

$$Z := \{(t,L) \in X \times Q;\ L \text{ is a jumping line of } E_t\}$$
$$(\text{resp. } Z := \{(t,C) \in X \times Q;\ C \text{ is a jumping conic of } E_t\})$$

has pure codimension 2 (resp. 3) and every component of Z maps generically finitely onto its image in X. In this case we call E regular (with respect to the given point(s)) and define D_2 as the image of Z under the projection to X.

3.4 **Lemma** : If E is regular the class of D_2 in Pic X is given by

$$[D_2] = n\, c_1(\mathcal{C}) - (n-1)\, c_1(\mathcal{B})$$
$$(\text{resp. } [D_2] = \tfrac{1}{2}(n-1)(n\, c_1(\mathcal{C}) - (n-2)\, c_1(\mathcal{A}))).$$

Proof : We consider the projections

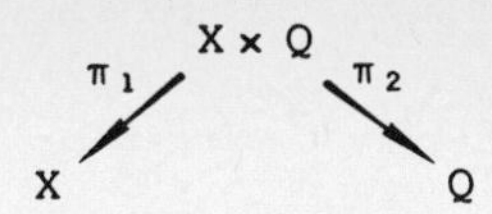

and define

$$\sigma := \pi_2^* \mathcal{O}_Q(1).$$

As above there are morphisms

$$\alpha : \pi_1^* \mathcal{B}(-\sigma) \longrightarrow \pi_1^* \mathcal{C} \quad (\text{resp. } \alpha: \pi_1^* \mathcal{A}(-\sigma) \longrightarrow \pi_1^* \mathcal{C})$$

and Z is given by

$$Z = \{(t,L) \in X \times Q;\ \text{rank } \alpha(t,L) \leq n-2\}$$

$$(\text{resp. } Z = \{(t,C) \in X \times Q;\ \text{rank } \alpha(t,C) \leq n-3\}).$$

If E is regular we can apply Porteous' formula ([1],p.86) to compute the class of Z. The result is

$$[Z] = \Delta_{2,1} (c_t(\pi_1^* \mathcal{B}(-\sigma) - \pi_1^* \mathcal{C}))$$

$$(\text{resp. } [Z] = -\Delta_{3,1}(c_t(\pi_1^* \mathcal{A}(-\sigma) - \pi_1^* \mathcal{C}))).$$

Since $[D_2] = \pi_{1*}[Z]$ it is sufficient to compute the terms of order 2 (resp.3) containing σ (resp.σ^2). A straightforward calculation gives the desired result.

The relation with Strømme's result is now obvious. The family E defines a map

$$\Phi : X \longrightarrow \overline{M}(c_1,c_2);\ t \longmapsto E_t .$$

If $c_1^2 - 4c_2 \not\equiv 0 \mod 8$ then

$$[D_1] = \Phi^*(\varepsilon) \quad (\text{resp.} [D_1] = \Phi^*(\varphi))$$

$$[D_2] = \Phi^*(\delta) \quad (\text{resp.} [D_2] = \Phi^*(\tfrac{1}{2}(n-1)\psi)).$$

Let us now return to the complete families of vector bundle which were constructed in section 1. I.e. let E be a vector bundle over $\mathbb{P}_2 \times \mathbb{P}_1$ given by an extension

$$0 \longrightarrow \mathcal{O}_X \to E(1,0) \longrightarrow \mathcal{I}_Y(a,b) \longrightarrow 0 .$$

Here (a,b) equals (1,2) or (2,1) and Y consists of k section $L_1,\ldots,L_k$. We set $c_1 = c_1(E_t) = a-2$ and $n = c_2(E_t) = k-a+1$.
As before we define

$$\mathcal{A} := R^1 p_{2*}E(-2,0),\ \mathcal{B} := R^1 p_{2*}E(-1,0),\ \mathcal{C} := R^1 p_{2*}E .$$

3.5 **Lemma : (i)** $c_1(\mathcal{A}) = 0$

(ii) $c_1(\mathcal{B}) = n$

(iii) $c_2(\mathcal{C}) = 2n-1 + c_1$.

Proof : Since $R^i p_{2*}E(-\ell,0)=0$ for $i \neq 1$ and $0 \leq \ell \leq 2$ the theorem of Grothendieck-Riemann-Roch yields

$$ch(-R^1 p_{2*}E(-\ell,0)) = p_{2*}[ch(E(-\ell,0)).\ Td(T_{\mathbb{P}_2})].$$

In order to do our computations it is useful to introduce the following notation :

$$\alpha := \mathcal{O}_X(1,0)\ ,\quad \beta := \mathcal{O}_X(0,1).$$

Then the Chow ring of X is generated by α and β modulo the relations $\alpha^3 = \beta^2 = 0$. Clearly the class of a section L_i is given by

$$L_i \sim \alpha^2 + \alpha\,\beta\ .$$

Hence we find

$$c_1(E(-\ell,0)) = [a-2(\ell+1)]\,\alpha + b\,\beta$$
$$c_2(E(-\ell,0)) = [k-a(\ell+1)+(\ell+1)^2]\,\alpha^2 + [k-b(\ell+1)]\,\alpha\beta\ .$$

A straightforward though somewhat tedious computation shows that the Chern character of $E(-\ell,0)$ is given by

$$\begin{aligned} ch(E(-\ell,0)) = {} & 2+[a-2(\ell+1)]\alpha + b\beta \\ & +[-k + \tfrac{a^2}{2} - a(\ell+1)+(\ell+1)^2]\,\alpha^2 \\ & +[ab-b(\ell+1)-k]\,\alpha\beta \\ & +[-\tfrac{k}{2} + k\ell + \tfrac{1}{2}a^2 b - ab(\ell+1) + \tfrac{b}{2}(\ell+1)^2]\,\alpha^2\beta. \end{aligned}$$

The Todd polynomial of $T_{\mathbb{P}_2}$ is

$$Td(T_{\mathbb{P}_2}) = 1 + \tfrac{3}{2}\alpha + \alpha^2\ .$$

The first Chern class of $R^1 p_{2*}E(-\ell,0)$ is then the negative coefficient of $\alpha^2\beta$ in the polynomial $ch(E(-\ell,0)).Td(T_{\mathbb{P}_2})$.

Therefore, we find

$$c_1(R^1 p_{2*}E(-\ell,0)) = 2k-k\ell-6+\tfrac{3}{2}\,b(\ell+1)+2(\ell+1)-\tfrac{b}{2}(\ell+1)^2.$$

From this our assertion follows readily.

We are now ready to state the main result of this section :

3.6 **Proposition :** Let E be a complete family of vector bundles as in section 1 with $c_1(E_t)=-1$ (resp.0) and $n=c_2(E_t)$. Then

(i) If the number of vector bundles E_t with a given jumping conic (jumping line) is finite, it is given by

$$\deg D_1 = 2n-2\ \ (\text{resp. } \deg D_1 = n).$$

(ii) If E is regular with respect to 1 (resp.3) given point(s) then the number of bundles with a jumping line (resp.jumping conic) through this (these) point(s) is

$$\deg D_2 = n(n-1)\ \ (\text{resp. } \deg D_2 = \tfrac{n}{2}(n-1)(2n-1)).$$

Proof : (i) Since the class of D_1 is given by $c_1(\mathcal{C})-c_1(\mathcal{A})$ (resp. $c_1(\mathcal{B})-c_1(\mathcal{A})$) this follows immediately from lemma 3.5.

(ii) By lemma 3.4 the class of D_2 is given by

$$\begin{aligned}[D_2] &= n\,c_1(\mathcal{C}) - (n-1)c_1(\mathcal{B}) \\ &= n(2n-2) - (n-1)n \\ &= n(n-1)\end{aligned}$$

$$\begin{aligned}(\text{resp.}[D_2] &= \frac{n-1}{2}(n\,c_1(\mathcal{C})-(n-2)\,c_1(\mathcal{A})) \\ &= \frac{n-1}{2}\,n(2n-1)).\end{aligned}$$

<u>Remark</u> : If $c_1=-1$ the number $\deg D_2$ can also be computed in a more naive way. To do this let P_o be a fixed point. Then E_t has a jumping line through P_o if and only if there are indices $i\neq j$ such that the points $\varphi_i(t),\varphi_j(t)$ and P_o are colinear. This is equivalent to

$$\det(\varphi_i(t),\ \varphi_j(t),P_o) = 0.$$

This determinant is of degree 2 with respect to t and since there are $\binom{n}{2}$ pairs of indices $i\neq j$ we find

$$\deg D_2 = 2\binom{n}{2} = n(n-1).$$

If

$$E_t|L = \mathcal{O}_L(-1-m) \oplus \mathcal{O}_L(m)$$

with $m\geq 1$ then m is called the <u>order</u> of the jumping line L. This is the case if and only if L contains m+1 of the points $\varphi_1(t),\ldots,\varphi_k(t)$. The above argument, therefore, shows that a jumping line of order m has to be counted with <u>multiplicity</u> $\binom{m+1}{2}$.

In the case $c_1=0$ there does not seem to exist a similarly easy way to compute $\deg D_2$ since the set of jumping conics of E_t does not necessarily depend on Y_t only.

<u>Remark</u> : Of course in the proof of 3.6 we could have deduced (ii) from (i) (or vice versa) using R.S=0. On the other hand we can now check our computations and recover R.S=0 as follows :

If $c_1=-1$ (resp.0) the class of S in Pic $\overline{M}(c_1,c_2)$ is

$$S\sim n\varepsilon - 2\delta \quad (\text{resp. } S\sim(2n-1)\varphi - \psi).$$

Let $\Phi: \mathbb{P}_1 \longrightarrow R\subseteq\overline{M}(c_1,c_2)$ be the induced map. Then

$$\deg \Phi^*(\varepsilon) = \deg D_1 = 2n-2, \qquad \deg \Phi^*(\delta) = \deg D_2 = n(n-1)$$

$$(\text{resp. } \deg \Phi^*(\varphi) = \deg D_1 = n\ , \qquad \deg \Phi^*(\psi) = \frac{2}{n-1}\ \deg D_2 = n(2n-1)).$$

This implies

$$\deg \Phi^*(S) = n(2n-2) - 2n(n-1) = 0$$

$$(\text{resp. } \deg \Phi^*(S) = (2n-1)n - n(2n-1) = 0)$$

and hence proves our claim.

4. FURTHER COMMENTS

Here we want to discuss a few <u>examples</u> :

If $c_1=-1$ and $c_2=1$ the moduli space M(-1,1) consists of only one point, namely $\overline{M}(-1,1) = M(-1,1) = \{\Omega_{\mathbb{P}_2}(1)\}$.

Hence there is nothing to say.

If $c_1=0$ and $c_2=2$ it was shown by Barth [3] that the moduli space $M(0,2)$ is isomorphic to

$$M(0,2)=\{\text{smooth conics in } \mathbb{P}_2\} \subseteq \mathbb{P}_5 = \mathbb{P}(\Gamma(\mathcal{O}_{\mathbb{P}_2}(2))).$$

Since the singular conics form a hypersurface $\Delta \subseteq \mathbb{P}_5$ the moduli scheme $M(0,2) \cong \mathbb{P}_5 \setminus \Delta$ is affine and we have

4.1 Proposition : There exist no complete curves in $M(0,2)$.

The next case is $c_1=-1$ and $c_2=2$. Then (cf. [6],p.256) we have :

$$M(-1,2) \cong \mathbb{P}_2 \times \mathbb{P}_2 \setminus \Delta_{/\Sigma_2}$$

where $\Delta \subseteq \mathbb{P}_2 \times \mathbb{P}_2$ is the diagonal and the symmetric group Σ_2 operates by interchanging the two factors $\mathbb{P}_2$. There exist complete curves in $M(-1,2)$. However

4.2 Proposition : There exist no complete surfaces in $M(-1,2)$.

Proof : Assume there is a complete surface $X \subseteq M(-1,2)$. Then there is also a complete surface $Y \subseteq \mathbb{P}_2 \times \mathbb{P}_2 - \Delta$, namely the inverse image of X under the quotient map. The Chow ring of $\mathbb{P}_2 \times \mathbb{P}_2$ is generated by $\alpha := \mathcal{O}(1,0)$ and $\beta := \mathcal{O}(0,1)$ modulo $\alpha^3 = \beta^3 = 0$. The class of Δ is

$$\Delta \sim \alpha^2 + \alpha\beta + \beta^2 .$$

The class of any surface Y is

$$Y \sim \ell\alpha^2 + m\alpha\beta + n\beta^2 \qquad (\ell,m,n \geq 0)$$

with at least one coefficient positive. Since

$$\Delta . Y = \ell + m + n > 0$$

we get a contradiction.

Already much more difficult is the case $c_1=0$ and $c_2=3$. There we have a map

$$J : \overline{M}(0,3) \longrightarrow \mathbb{P}_9 = \mathbb{P}(\Gamma(\mathcal{O}_{\mathbb{P}_2}(3)))$$
$$E \longmapsto J(E)$$

which associates to each sheaf E its curve of jumping lines. Restricted to $M(0,3)$ the map J has finite fibres and is of degree 3. The image of the boundary S under the map J is the variety $\overline{S} \subseteq \mathbb{P}_9$ of reducible cubics. $\overline{S}$ is a 7-dimensional variety of degree 21. The inverse image of a curve consisting of a smooth conic and a transversal line consists of precisely one stable bundle and a $\mathbb{P}_1$ worth of stable sheaves which are not locally free. There are no vector bundles lying over a cubic which is the union of a conic and a tangential line. For a discussion of the map J see Barth [3] and Maruyama [9]. This implies immediately that $M(0,3)$ does not contain a complete threefold. The existence of a complete surface $X \subseteq M(0,3)$ is a more difficult question. Its image $Y:=J(X)$ under J would necessarily intersect the variety $\overline{S}$ of reducible cubics. Hence to find a surface $X \subseteq M(0,3)$ amounts to finding a surface $Y \subseteq M(0,3)$ whose inverse image under J decomposes into at least two components one of which does not intersect the boundary S. We were unfortunately not able to answer the question whether such a surface exists or not.

REFERENCES

[1] **Arbarello,E., M.Cornalba, P.A.Griffiths , J.Harris,** Geometry of Algebraic Curves, Vol.I, Grundlehren 267, Springer Verlag, 1984

[2] **Barth,W.,** Some Properties of Stable Rank 2 Vector Bundles on $\mathbb{P}_n$. Math.Ann. 226, 125-150 (1977)

[3] **Barth,W.,** Moduli of Vector Bundles on the Projective Plane. Inv. Math. 42, 63-91 (1977)

[4] **Brun,J.,** Les fibrés de rang deux sur $\mathbb{P}_2$ et leurs sections. Bull. Soc. Math. France 108, 457-473 (1980)

[5] **Brun,J., A.Hirschowitz,** Restrictions planes du fibré instanton général, Forthcoming

[6] **Hulek,K.,** Stable Rank 2 Vector Bundles on $\mathbb{P}_2$ with c_1 Odd. Math. Ann.242, 241-266 (1979)

[7] **Le Potier,J.,** Fibrés stables de rang 2 sur $\mathbb{P}_2(\mathbb{C})$. Math.Ann. 241, 217-256 (1979)

[8] **Maruyama,M.,** Moduli of Stable sheaves,II. J.Math. Kyoto Univ. 18, 557-614 (1978)

[9] **Maruyama,M.,** Singularities of the Curve of Jumping Lines of a Vector Bundle of Rank 2 on $\mathbb{P}^2$. in: Algebraic Geometry, Springer LN 1016, 370-411 (1984)

[10] **Okonek,C., M.Schneider, H.Spindler,** Vector Bundles on Complex Projective Spaces, Progress in Math. vol.3, Birkhaüser, Boston (1980)

[11] **Strømme,S.A.,** Ample Divisors on Fine Moduli Spaces on the Projective Plane. Math. Z. 187, 405-423 (1984).

*

Appendix to the paper "Complete Families of Stable Vector Bundles over $\mathbb{P}_2$".

K. Hulek and S. A. Strømme

In this appendix we want to describe two more techniques which enable us to construct complete families of stable vector bundles. The first method uses essentially the fact that the boundary of the moduli space $\bar{M}_{\mathbb{P}_2}(c_1,c_2)$ can be "blown down" to a variety of codimension two. The other method uses stable vector bundles on higher dimensional projective spaces. In some cases this also allows us to construct examples of complete surfaces of stable vector bundles over $\mathbb{P}_2$.

I. Here the key point is the following:

Proposition: Let $c_1 = 0$ (resp. $c_1 = -1$) and $n = c_2 \geq 3$ (resp. $n = c_2 \geq 2$). The map

$$\phi : \bar{M}_{\mathbb{P}_2}(c_1,c_2) \to \mathbb{P}_N := \mathbb{P}(\Gamma(\mathcal{O}_{\mathbb{P}_2}(n))) \quad (\text{resp. } \mathbb{P}(\Gamma(\mathcal{O}_{\mathbb{P}_5}(n-1))))$$

$$F \mapsto J(F)$$

which associates to each sheaf F its divisor of jumping lines (resp. jumping conics) is a morphism. It is generically finite and maps the boundary $S = \bar{M}_{\mathbb{P}_2}(c_1,c_2) \smallsetminus M_{\mathbb{P}_2}(c_1,c_2)$ to a variety of codimension 2 in the image of $\bar{M}_{\mathbb{P}_2}(c_1,c_2)$.

Proof: We shall restrict ourselves to the case $c_1 = 0$. The case $c_1 = -1$ is analogous. We first remark that the map $F \mapsto J(F)$ is well defined, i.e. that not every line is a jumping line. For vector bundles this is the Grauert-Mülich theorem. For semi-stable sheaves F this can be seen by applying the Grauert-Mülich theorem to the double dual $F^{\vee\vee}$ of F. That the map $F \mapsto J(F)$ is a morphism follows from [8],p.377 and [9]p.418. The assertion that ϕ is generically finite follows from the fact that a smooth curve has only finitely many different theta-characteristics together with Barth's classification result [2]p.73.

Since S is irreducible ([9]p.406 and [6]p.57) it remains to show that the dimension of a general fibre of the map $\phi|S : S \to \phi(S)$ is one. To see this, let $F' \in M_{\mathbb{P}_2}(0,n-1)$ be a stable rank 2 vector bundle. Let $P \in \mathbb{P}_2$ be an arbitrary point. For each choice of a surjection

$$\lambda : F'_P \to \mathbb{C}_P \to 0$$

we get an exact sequence

$$0 \to F \to F' \overset{\lambda}{\to} \mathbb{C}_P \to 0$$

where F is a semi-stable sheaf with $c_1(F) = 0$ and $c_2(F) = n$. Its curve of jumping lines is

$$J(F) = J(F') \cup L_P$$

where $L_p \subseteq \mathbb{P}_2^*$ denotes the line parametrizing all lines $L \subseteq \mathbb{P}_2$ through P. Hence $J(F)$ does not depend on the choice of λ. The isomorphism class of F, however, does depend on

$$\lambda \in \mathbb{P}_1 = \mathbb{P}(\mathrm{Hom}(F'_P, \mathbb{C}_P)).$$

This gives a family $\mathbb{P}_1 \subseteq \bar{M}_{\mathbb{P}_2}(c_1,c_2)$ of semi-stable sheaves which under ϕ are all mapped to the same point $J(F)$. Varying the bundle $F' \in M_{\mathbb{P}_2}(0,n-1)$, the point $P \in \mathbb{P}_2$ and the surjection $\lambda \in \mathbb{P}_1$, we get a family of dimension

$$\begin{aligned}\dim M_{\mathbb{P}_2}(0,n-1) + 2 + 1 &= 4(n-1) - 3 + 3 = 4n - 4\\ &= \dim S\end{aligned}$$

i.e. an open dense subset of S. This completes the proof.

<u>Corollary:</u> <u>Let</u> $c_1 = 0$ (<u>resp.</u> $c_1 = -1$) <u>and</u> $n = c_2 \geq 3$ (<u>resp.</u> $n = c_2 \geq 2$). <u>Then there exist complete curves</u> $C \subseteq M_{\mathbb{P}_2}(c_1,c_2)$ <u>through a general point of</u> $M_{\mathbb{P}_2}(c_1,c_2)$.

<u>Proof:</u> Intersect $\phi(\bar{M}_{\mathbb{P}_2}(c_1,c_2))$ with a general linear subspace $\mathbb{P}_m \subseteq \mathbb{P}_N$ of dimension

$$m = N - \dim \bar{M}_{\mathbb{P}_2}(c_1,c_2) + 1.$$

II. Here we use stable bundles on higher dimensioal projective spaces in order to construct complete families of stable vector bundles over $\mathbb{P}_2$.

We first recall that the "general" instanton bundle $\bar{F}$ on $\mathbb{P}_3$ (for details see [4]) for $c_2 \geq 5$ has no jumping planes. J.e. the restriction $\bar{F}|E$ for every pane $E \subseteq \mathbb{P}_3$ is stable.

Proposition: Let F be a rank 2 vector bundle on $\mathbb{P}_2$ with $c_1 = 0$ and $c_2 \geq 5$ which can be extended to a general instanton bundle on $\mathbb{P}_3$. Then a complete rational curve $R \subseteq M_{\mathbb{P}_2}(0,c_2)$ through the point $[F] \in M_{\mathbb{P}_2}(0,c_2)$ exists.

Proof: Let $\bar{F}$ be an extension of F to $\mathbb{P}_3$ which is a general instanton bundle. We consider the Segre-embedding

$$s : \mathbb{P}_1 \times \mathbb{P}_2 \to \mathbb{P}_5 .$$

Let $\Lambda \subseteq \mathbb{P}_5$ be a general line. Projection from Λ gives a 3:1 map

$$\pi_\Lambda : \mathbb{P}_1 \times \mathbb{P}_2 \to \mathbb{P}_3 .$$

We set

$$\tilde{F} := \pi_\Lambda^* \bar{F} .$$

Since $\bar{F}$ has no jumping planes we can consider $\tilde{F}$ as a complete family of stable vector bundles over $\mathbb{P}_2$ with parametrizing space $\mathbb{P}_1$. It remains to prove that the map

$$\varphi : \mathbb{P}_1 \to M_{\mathbb{P}_2}(0,c_2)$$
$$t \mapsto \tilde{F}|\{t\} \times \mathbb{P}_2$$

is not constant. To see this we look at the projections

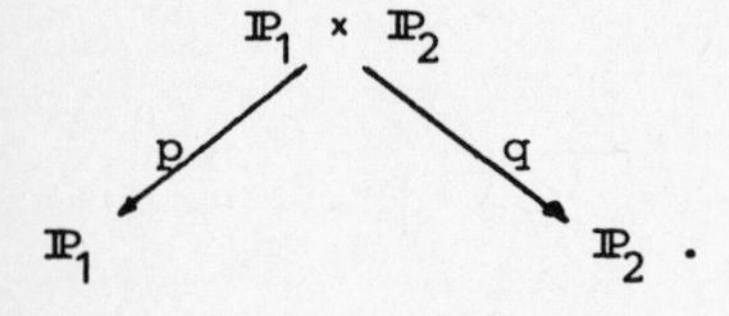

Set

$$\alpha := p^*\mathcal{O}_{\mathbb{P}_1}(1), \quad \beta := q^*\mathcal{O}_{\mathbb{P}_2}(1)$$

Then by construction

$$c(\tilde{F}) = 1 + n(\alpha+\beta)^2 = 1 + 2n\alpha\beta + n\beta^2.$$

If φ were constant this would imply that

$$\tilde{F} = q^* F \otimes p^*\mathcal{O}_{\mathbb{P}_1}(k)$$

for some $k \in \mathbb{Z}$. But then

$$c(\tilde{F}) = 1 + 2k\alpha + n\beta^2 .$$

This implies $k = 0$ and contradicts the above expression for $c(\tilde{F})$.

Remark: This gives another proof of the corollary in section I, at least for $c_1 = 0$ and $c_2 \geq 5$, since the general rank 2 bundle on $\mathbb{P}_2$ can be extended to an instanton bundle on $\mathbb{P}_3$ (see [5]).

Finally we want to use the Horrocks-Mumford bundle F on $\mathbb{P}_4$ to construct complete surfaces of stable rank 2 bundles over $\mathbb{P}_2$.

Proposition: There exist complete rational surfaces $X \subseteq M_{\mathbb{P}_2}(-1,4)$.

Proof: Recall that the variety of jumping planes $S(F) \subseteq Gr(2,4)$ of the Horrock-Mumford bundle is a surface [3]p.5. Let

$$s : \mathbb{P}_2 \times \mathbb{P}_2 \to \mathbb{P}_9$$

be the Segre-embedding and choose a general linear subspace $\Lambda \subseteq \mathbb{P}_9$ of dimension 4 which does not intersect $s(\mathbb{P}_2 \times \mathbb{P}_2)$. Projection from Λ gives a 6:1 map

$$\pi_\Lambda : \mathbb{P}_2 \times \mathbb{P}_2 \to \mathbb{P}_4 .$$

As before we set

$$\tilde{F} := \pi_\Lambda^* F .$$

F can be viewed as a family of rank 2 vector bundles over $\mathbb{P}_2$ with base space $\mathbb{P}_2$. We first have to see that we can find a Λ such that for each $t \in \mathbb{P}_2$ the bundle

$$\tilde{F}_t := \tilde{F} | \{t\} \times \mathbb{P}_2$$

is stable. Let

$$\bar{\pi}_\wedge : \mathbb{P}_2 \to Gr(2,4)$$

be the map which associates to each $t \in \mathbb{P}_2$ the plane $\pi_\wedge(\{t\} \times \mathbb{P}_2) \subseteq \mathbb{P}_4$. We have to see that

$$\bar{\pi}_\wedge(\mathbb{P}_2) \cap S(F) = \emptyset .$$

But by Kleiman's transversality theorem [7] we can find a transformation $g \in PGL(5,\mathbb{C})$ such that

$$\bar{\pi}_\wedge(\mathbb{P}_2) \cap g^{-1}(S(F)) = \bar{\pi}_\wedge(\mathbb{P}_2) \cap S(g^*F) = \emptyset .$$

Hence after possibly replacing F by g^*F we can assume that $\tilde{F}$ is a family of stable vector bundles.
It remains to show that the map

$$\varphi : \mathbb{P}_2 \to M_{\mathbb{P}_2}(-1,4)$$
$$t \mapsto \tilde{F} | \{t\} \times \mathbb{P}_2$$

is not constant. Just as in the proof of the previous proposition this can be done by computing the Chern classes of $\tilde{F}$.

<u>Corollary:</u> <u>Let</u> $\bar{d}$ <u>be an integer</u> ≥ 1. <u>Then the moduli spaces</u> $M_{\mathbb{P}_2}(0,15\bar{d}^2)$ <u>and</u> $M_{\mathbb{P}_2}(-1,15\bar{d}(\bar{d}+1)+4)$ <u>contain complete rational surfaces.</u>

<u>Proof:</u> Let $h_o, h_1, h_2 \in \Gamma(\mathcal{O}_{\mathbb{P}_2}(d))$ be homogeneous forms of degree d without common zeroes. They define a branched covering

$$h : \mathbb{P}_2 \to \mathbb{P}_2$$

of degree d^2. Let $\tilde{F}$ be the rank 2 bundle over $\mathbb{P}_2 \times \mathbb{P}_2$ which was constructed in the proof of the preceding proposition. We consider the map

$$\tilde{h} := (\mathrm{id},h) : \mathbb{P}_2 \times \mathbb{P}_2 \to \mathbb{P}_2 \times \mathbb{P}_2$$

and set

$$\tilde{G} := \tilde{h}^*(\tilde{F}).$$

Since the pull-back of a stable bundle remains stable [1] p.133 we can consider $\tilde{G}$ again as a family of stable rank 2 vector bundles over $\mathbb{P}_2$ with base space $\mathbb{P}_2$. Clearly $\tilde{G}$ is non-degenerate. It remains to compute the Chern classes of the bundles $\tilde{G}_t$.

To do this let E be any rank 2 vector bundle on $\mathbb{P}_2$ with Chern classes $c_1(E) = c_1$ and $c_2(E) = c_2$. We set

$$\Delta(E) = 4c_2 - c_1^2 .$$

Note that

$$\Delta(E(k)) = \Delta(E)$$

for all $k \in \mathbb{Z}$. Since

$$\Delta(h^*E) = d^2\Delta(E)$$

we find

$$\Delta(\tilde{G}_t) = d^2\Delta(\tilde{F}_t) = 15d^2.$$

Now assume that d is even, say $d = 2\bar{d}$. Then

$$\Delta(\tilde{G}_t) = 60\bar{d}^2.$$

It follows that $c_1(\tilde{G}_t)$ is even and after normalizing it to $c_1 = 0$ one finds

$$4c_2(\tilde{G}_t^{norm}) = 60\bar{d}^2$$

hence

$$c_2(\tilde{G}_t^{norm}) = 15\bar{d}^2 .$$

The case d odd is treated in the same way.

References

[1] Barth, W.: Some properties of stable rank-2 vector bundles on $\mathbb{P}_n$. Math. Ann. 226, 125-150 (1977).

[2] Barth, W.: Moduli of vector bundles on the projective plane. Inv. Math. 42, 63-91 (1977).

[3] Barth, W., Hulek, K., Moore, R.: Shioda's modular surface S(5) and the Horrocks-Mumford bundle. To appear in: Proceedings of the International Conference on Algebraic Vector Bundles over Algebraic Varieties. Bombay 1984.

[4] Brun, J., Hirschowitz, A.: Restrictions planes du fibré instanton général. Forthcoming.

[5] Donaldson, S.K.: Instantons and geometric invariant theory. Commun. Math. Phys. 93, 453-460 (1984).

[6] Drezet, J.-M.: Groupe de Picard des variétés de modules de faisceaux semi-stable sur $\mathbb{P}_2(\mathbb{C})$. To appear.

[7] Kleiman, S.: The transversality of a general translate. Comp. Math. 128, 287-297 (1974).

[8] Maruyama, M.: Singularities of the curves of jumping lines of a vector bundle of rank 2 on $\mathbb{P}^2$. Springer Lecture Notes, Vol. 1016, 370-411.

[9] Strømme, S.A.: Ample divisors on fine moduli spaces on the projective plane. Math. Z. 187, 405-423 (1984).

K. Hulek
Mathematisches Institut
Universität Bayreuth
Postfach 3oo8
858o Bayreuth
West Germany

S. A. Strømme
Dept. of Mathematics
Bergen University
N - 5o14 Bergen
Norway

ON THE MINIMAL MODEL PROBLEM

Yujiro Kawamata
Department of Mathematics, University of Tokyo
Hongo, Tokyo, 113, Japan

§ 1. Introduction

Let X be a non-singular projective variety defined over $k = \mathbb{C}$. We define the *canonical ring* $R(X)$ by

$$R(X) = \bigoplus_{m \geqq 0} H^0(X, mK_X)$$

with a natural graded k-algebra structure. The *Kodaira dimension* $\kappa(X)$ ([I], [U]) is defined as

$$\kappa(X) = \begin{cases} \text{trans.deg.}_k R(X) - 1 & \text{if } R(X) \neq k \\ -\infty & \text{otherwise .} \end{cases}$$

X is said to be of *general type* if $\kappa(X) = \dim X$. Let us consider the following problem (cf. [W1]) :

PROBLEM. *Is $R(X)$ finitely generated over k ?*

We can also consider a relative version of this problem : let $\mu : X \longrightarrow S$ be a desingularization of some algebraic variety S and consider the *relative canonical ring*

$$\mathcal{R}(X/S) = \bigoplus_{m \geqq 0} \mu_* \mathcal{O}_X(mK_X).$$

PROBLEM'. *Is $\mathcal{R}(X/S)$ finitely generated as a graded $\mathcal{O}_S$-algebra ?*

The answers to the above problems are affirmative in case $\dim X = 2$. Let us recall the proof. A curve C on a non-singular surface X is called a (-1)- (resp. (-2)-) *curve* if $C \simeq \mathbb{P}^1$ and $(C^2) = -1$ (resp. -2). The contraction criterion of Castelnuovo and Enriques says that a (-1)-curve can be contracted to a point of another non-singular surface X' ; the morphism $X \longrightarrow X'$ is the inverse of the blowing-up at that point. Starting from a non-singular projective surface and repeating this contraction process, we obtain a non-singular projective surface X_{min} without (-1)-curves, i.e., a *minimal model*. If $\varkappa(X) = 2$, then by Artin's contraction theorem ([A]), we can contract all the (-2)-curves on X_{min} ; there is a morphism $X_{min} \longrightarrow X_{can}$ to the *canonical model* X_{can} , which is a normal projective surface having at most rational double points. Then the canonical divisor $K_{X_{can}}$ is ample, and hence $R(X) \simeq R(X_{can})$ is finitely generated ([Mf], cf. [K], [Bo]). The second problem is solved similarly.

The minimal model thus obtained is also the starting point of the Kodaira-Enriques classification of surfaces. There are two basic theorems :

(1) *X_{min} is ruled or $\mathbb{P}^2$ if $\varkappa(X) = -\infty$.*

(2) *In case $\varkappa(X) > 0$, there is a positive integer m such that $|mK_{min}|$ is base point free.*

Thus $K_{X_{min}}$ is *nef* (i.e., $(K_{X_{min}}.C) \geqq 0$ for all curves C) unless X is birationally ruled.

It was believed impossible to extend the minimal model theory to higher dimensional cases. For example, there is a counterexample for the existence of the minimal model by Francia [F] ; which is now a good example of a flip (see also [Fk]). But Mori [Mo1] proved a contraction theorem for non-singular projective 3-folds by introducing the concept of extremal rays which is a generalization of (- 1)-curves. His contraction morphism produces singularities and one cannot proceed inductively to reach minimal models. This phenomenon was anticipated by Reid's works [R1], [R2]. In fact, Reid set up a category of varieties which should be the framework for the inductive procedure ; we look for minimal models in the category of $\mathbb{Q}$-factorial terminal projective varieties. In this paper we shall explain what are proved and what are still conjectures in the minimal model problem. Main results by now obtained are the contraction theorem and the cone theorem. We have yet to prove the flip conjectures. We refer the reader to [KMM] for a more detailed survey with proofs.

§ 2. The minimal model conjecture

We fix our terminology. The base field k is always assumed to be $\mathbb{C}$. Let X be a normal algebraic variety of dimension d. The group of Weil (resp. Cartier) divisors on X are denoted by $Z_{d-1}(X)$ (resp. $\mathrm{Div}(X)$). The *canonical divisor* K_X is the image of $K_{\mathrm{Reg}(X)}$ by an isomorphism $Z_{d-1}(\mathrm{Reg}(X)) \longrightarrow Z_{d-1}(X)$, where $\mathrm{Reg}(X)$ is the regular part of X. For $D \in Z_{d-1}(X)$, $\mathcal{O}_X(D)$ denotes the corresponding reflexive sheaf of rank 1. Thus $\mathcal{O}_X(K_X) = \omega_X$. The elements of $Z_{d-1}(X)_{\mathbb{Q}}$ (resp. $\mathrm{Div}(X)_{\mathbb{Q}}$) are called $\mathbb{Q}$-*divisors* (resp. $\mathbb{Q}$-*Cartier divisors*), where the subscript $\mathbb{Q}$ means $\otimes \mathbb{Q}$. $D = \Sigma a_j D_j \in Z_{d-1}(X)_{\mathbb{Q}}$ being the decomposition by irreducible components, we define the *integral part* of D by $[D] = \Sigma [a_j] D_j$. X is said to be $\mathbb{Q}$-*factorial* if the natural injective homomorphism $\mathrm{Div}(X)_{\mathbb{Q}} \longrightarrow Z_{d-1}(X)_{\mathbb{Q}}$ is also surjective. A birational morphism $f : X \longrightarrow X'$ is called an *isomorphism in codimension 1* if it induces an isomorphism $f_* : Z_{d-1}(X) \longrightarrow Z_{d-1}(X')$, where $d = \dim X$. The existence of isomorphisms in codimension 1 adds new dimension of complexity for algebraic varieties of dimension $\geqq 3$.

The singularity of a normal algebraic variety is called *terminal* (resp. *canonical*) if the following conditions are satisfied :

(i) K_X is $\mathbb{Q}$-Cartier, and

(ii) given a desingularization $\mu : Y \longrightarrow X$ which has a simple normal crossing exceptional locus $\cup F_j$ ([H]), we can write

$$K_Y = \mu^* K_X + \Sigma a_j F_j \quad \text{in} \quad \mathrm{Div}(Y)_{\mathbb{Q}}$$

with $a_j > 0$ (resp. $a_j \geqq 0$) for all j.

The a_j are called the *discrepancies*, and the smallest positive integer r such that $rK_X \in \mathrm{Div}(X)$ is called the *index*. If X is affine and $\mathcal{O}_X(rK_X)$ is trivial, then there is an r-fold cyclic covering $Y \longrightarrow X$ which is etale in codimension 2 and such that Y has terminal (resp. canonical) singularities of index 1. This is called the *canonical cover* of X.

More generally, a pair (X, Δ) of a normal algebraic variety and a $\mathbb{Q}$-divisor is called *log-terminal* if

(i) $[\Delta] = 0$ and $K_X + \Delta \in \mathrm{Div}(X)_{\mathbb{Q}}$, and

(ii) given a desingularization $\mu : Y \longrightarrow X$ such that the union $\cup F_j$ of the exceptional locus and the inverse image of the support of Δ is a simple normal crossing divisor, we can write

$$K_Y = \mu^*(K_X + \Delta) + \Sigma a_j F_j \quad \text{in} \quad \mathrm{Div}(X)_{\mathbb{Q}}$$

with $a_j > -1$ for all j.

The a_j are called the *log-discrepancies*. If (X, Δ) is log-terminal, then X has only rational singularities ([E], [Ft2], [KMM]). By introducing the "log" terminology, we can deal with (-2)-curves in the same routine as (-1)-curves. In general, if X has at most quotient singularities, then $(X, 0)$ is log-terminal. If (X, Δ) is log-terminal and Δ' is an effective Cartier divisor on X, then $(X, \Delta + \varepsilon\Delta')$ is log-terminal for a rational number ε such that $0 < \varepsilon << 1$. For example, if $\dim X = 2$, then

terminal $\longleftrightarrow$ non-singular

canonical $\longleftrightarrow$ rational double points

log-terminal with $\Delta = 0$ $\longleftrightarrow$ quotient singularities.

Let $f : X \longrightarrow S$ be a projective surjective morphism of normal algebraic varieties. Let $Z_1(X/S)$ be the free abelian group generated by irreducible curves which are mapped to points by f. The intersection pairing gives a bilinear map $\mathrm{Div}(X) \times Z_1(X/S) \longrightarrow \mathbb{Z}$, and the *numerical equivalence* $\approx$ is defined so that the pairing $(\mathrm{Div}(X)/\approx)_{\mathbb{Q}} \times (Z_1(X/S)/\approx)_{\mathbb{Q}} \longrightarrow \mathbb{Q}$ is non-degenerate. We define

$N^1(X/S) = (\mathrm{Div}(X)/\approx)_{\mathbb{R}}$, $N_1(X/S) = (Z_1(X/S)/\approx)_{\mathbb{R}}$

$\overline{NE}(X/S)$ = the closed convex cone in $N^1(X/S)$ generated by effective 1-cycles

$$\rho(X/S) = \dim_{\mathbb{R}} N^1(X/S).$$

$D \in \mathrm{Div}(X)_{\mathbb{Q}}$ is said to be *f-ample* (resp. *f-nef*) if the linear functional defined by D on $N^1(X/S)$ is positive (resp. non-negative) on $\overline{NE}(X/S) - \{0\}$ ([Kl]). D is called *f-big* if $\varkappa(X_\eta, D_\eta) = \dim X_\eta$ for the generic point η of S, where $\varkappa$ denotes Iitaka's D-dimension. In this case there is an f-ample $\mathbb{Q}$-Cartier divisor A and an effective $\mathbb{Q}$-Cartier divisor B such that $D = A + B$ by Kodaira's lemma. The morphism $f : X \longrightarrow S$ is called *minimal* (resp. *canonical*) if X has at most canonical singularities and if K_X is f-nef (resp. f-ample). In case S is a point, X is called a *minimal* (resp. *canonical*) *variety*. Then our problem is formulated as follows :

THE MINIMAL MODEL CONJECTURE. *Let $f : X \longrightarrow S$ be a projective surjective morphism of algebraic varieties. Assume that the generic fiber X_η is not uniruled. Then there is a minimal $f' : X' \longrightarrow S$ which is birationally equivalent to f over S.*

The problems in the introduction are reduced to the above conjecture by the following theorem, which is an easy corollary of the contraction theorem in § 3.

THEOREM 1. *Let $f : X \longrightarrow S$ be a minimal morphism and assume that K_X is f-big. Then*

$$\mathcal{R}(X/S) = \bigoplus_{m \geqq 0} f_* \mathcal{O}_X(mK_X)$$

is a finitely generated graded $\mathcal{O}_S$-algebra. Thus the natural morphism $\mathrm{Proj}\, \mathcal{R}(X/S) \longrightarrow S$ is canonical.

We write $f : (X, \Delta) \longrightarrow S$ for the pair of a morphism $f : X \longrightarrow S$

as above and a $\mathbb{Q}$-divisor Δ on X. It is called *log-minimal* if the pair (X, Δ) is log-terminal and $K_X + \Delta$ is f-nef.

THE LOG-MINIMAL MODEL CONJECTURE. *Let $f : X \longrightarrow S$ be a projective surjective morphism from a non-singular variety X to a normal variety S, and let Δ be a $\mathbb{Q}$-divisor on X with $[\Delta] = 0$ whose support is simple normal crossing. Assume that $\kappa(X_\eta, K_{X_\eta} + \Delta_\eta) \geqq 0$ for the generic fiber X_η. Then there is a birational model $f' : (X', \Delta') \longrightarrow S$ of f which is log-minimal with $\Delta' \in Z_{d-1}(X')$ being the strict transform of Δ.*

THEOREM 2. *Let $f : (X, \Delta) \longrightarrow S$ be a log-minimal morphism and assume that $K_X + \Delta$ is f-big. Then*

$$\mathcal{R}(X/S, K_X + \Delta) = \bigoplus_{m \geqq 0} f_* \mathcal{O}_X([m(K_X + \Delta)])$$

is finitely generated.

§ 3. A minimal model program

We have the following two basic theorems toward the construction of (log-)minimal models. These are obtained as combinations of works by many authors ([Mo1], [Re2,3], [Ka2,3,4], [B], [S1,2], [Ko], [KMM]).

THE CONTRACTION THEOREM. *Let $f : X \longrightarrow S$ be a projective surjective morphism of normal algebraic varieties, Δ a $\mathbb{Q}$-divisor on X such that (X, Δ) is log-terminal, and let D be a Cartier divisor on X. Assume that D is f-nef and that $D - (K_X + \Delta)$ is* *f-nef and f-big. Then there exists a positive integer* m_0 *such that the natural homomorphism* $f^*f_*\mathcal{O}_X(mD) \longrightarrow \mathcal{O}_X(mD)$ *is surjective for any* $m \geqq m_0$.

Remark. The contraction theorem is also true in the case where X is a complex manifold and f is a proper surjective morphism which is bimeromorphically equivalent to a projective surjective morphism.

THE CONE THEOREM. *Let* $f : X \longrightarrow S$ *be a projective surjective morphism of normal algebraic varieties and let* Δ *be a* $\mathbb{Q}$*-divisor on* X *such that* (X, Δ) *is log-terminal. Then* $\overline{NE}(X/S)$ *is locally rational polyhedral in the half space* $N^1(X/S)^- = \left\{ z \in N^1(X/S) \ ; \ ((K_X + \Delta).z) < 0 \right\}$:

$$\overline{NE}(X/S) = \overline{NE}(X/S)^+ + \Sigma R_i ,$$

where $\overline{NE}(X/S)^+ = \left\{ z \in \overline{NE}(X/S) \ ; \ ((K_X + \Delta).z) \geqq 0 \right\}$ *and the* R_i *are rational extremal rays which are discrete in the half space* $N^1(X/S)^-$.

These theorems are obtained as applications of the following vanishing theorem ([Ka1], [V], [KMM]) :

THEOREM 3. *Let* $f : X \longrightarrow S$ *be a projective surjective morphism of normal algebraic varieties,* Δ *a* $\mathbb{Q}$*-divisor on* X*, and let* D *be a Weil divisor on* X *which is* $\mathbb{Q}$*-Cartier. Assume that* (X, Δ) *is log-terminal and that* $D - (K_X + \Delta)$ *is f-nef and f-big. Then*

$$R^i f_* \mathcal{O}_X(D) = 0 \quad \text{for} \quad i > 0.$$

The (log-)minimal model problem will be solved if the folllowing (log-)flip conjectures are affirmative.

THE LOG-FLIP CONJECTURE I (the existence of the log-flip). *Let* (X, Δ) *be a log-terminal pair such that* X *is* $\mathbb{Q}$*-factorial and let* $\varphi : X \longrightarrow Z$ *be a projective surjective morphism which is an isomorphism in codimension 1. Assume that* $\rho(X/Z) = 1$ *and that* $-(K_X + \Delta)$ *is* φ*-ample. Then there exists a projective surjective morphism* $\varphi' : X' \longrightarrow Z$ *from a* $\mathbb{Q}$*-factorial variety such that,* Δ' *being the strict transform of* Δ*, the pair* (X', Δ') *is log-terminal,* $\rho(X'/Z) = 1$*, and such that* $K_{X'} + \Delta'$ *is* φ'*-ample.*

THE LOG-FLIP CONJECTURE II (the termination of log-flips). *There does not exist an infinite chain of log-flips consisting of relatively projective varieties over a fixed variety* S *as follows.*

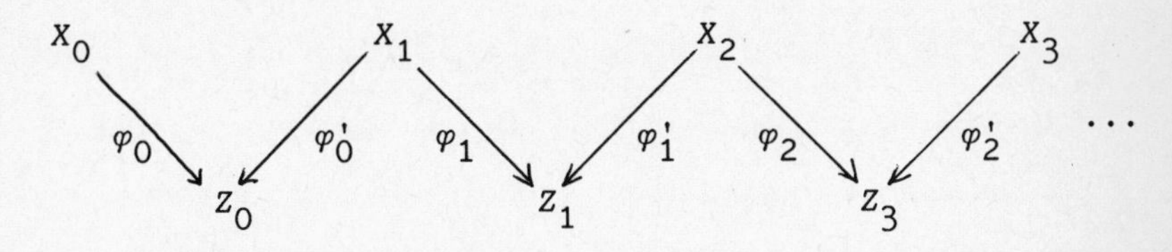

The *flip conjectures* are special cases of the log-flip conjectures where $\Delta = 0$ and X has terminal singularities.

Let $f_0 : X_0 \longrightarrow S$ be a projective surjective morphism from a non-singular variety to a normal variety and let Δ_0 be a $\mathbb{Q}$-divisor on X_0 with $[\Delta_0] = 0$ whose support is a simple normal crossing divisor. Assume that $\kappa(X_{0,\eta}, K_{X_{0,\eta}} + \Delta_{0,\eta}) \geqq 0$. We shall explain how to obtain a log-minimal model of f_0 assuming the log-flip conjectures. We construct inductively the following sequence of morphisms $f_n : X_n \longrightarrow S$ from log-terminal pairs (X_n, Δ_n) such that the X_n are $\mathbb{Q}$-factorial and which are birationally equivalent to f_0. If $K_{X_n} + \Delta_n$ is f_n-nef, then it is the desired log-minimal model. If not, there is an extremal ray

R of $\overline{NE}(X_n/S)$ by the cone theorem. Let D be a Cartier divisor on X_n which supports R ; i.e., D is f_n-nef and

$$\left\{ z \in \overline{NE}(X_n/S) \ ; \ (D.z) = 0 \right\} = R.$$

Then by the contraction theorem, some positive multiple mD gives a projective surjective morphism $\varphi : X_n \longrightarrow Z$ to a normal relatively projective variety Z over S with connected fibers such that $-(K_{X_n} + \Delta_n)$ is φ-ample. Since $\varkappa(X_{n,\eta}, K_{X_{n,\eta}} + \Delta_{n,\eta}) \geqq 0$, φ is a birational morphism. There are two cases.

(i) φ is *divisorial* ; φ contracts a prime divisor on X_n . We set $X_{n+1} = Z$ and $\Delta_{n+1} = \varphi_* \Delta_n$, the strict transform. Then $\rho(X_{n+1}/S) = \rho(X_n/S) - 1$.

(ii) φ is *flipping* ; φ is an isomorphism in codimension 1. By the log-flip conjecture I, we obtain $\varphi' : X_{n+1} \longrightarrow Z$. We set Δ_{n+1} the strict transform of Δ_n . In any case, this sequence will terminate after a finite number of steps and yields the log-minimal model.

If we start with a terminal X_n , then X_{n+1} is also terminal. By [MM], if $X_{0,\eta}$ is not uniruled, then φ is a birational morphism. Thus the flip conjectures imply the minimal model conjecture similarly. Actually, we obtain here a strictly (log-)terminal, i.e., $\mathbb{Q}$-factorial (log-)terminal, minimal model in this procedure.

§ 4. Some results

The following conjecture is essential in classifying minimal algebraic varieties.

THE ABUNDANCE CONJECTURE. *Let* X *be a minimal algebraic variety. Then*

$$\varkappa(X) = \nu(X) =_{\mathrm{def}} \max\left\{ k \;;\; (K_X)^k \not\approx 0 \right\}.$$

THEOREM 4 ([Ka5]). *Let* $f : X \longrightarrow S$ *be a minimal morphism. Assume that* $\varkappa(X_\eta) = \nu(X_\eta)$ *for the generic fiber* X_η *of* f. *Then* f *is good, i.e., there is a positive integer* m *with* $mK_X \in Div(X)$ *such that the natural homomorphism* $f^*f_*\mathcal{O}_X(mK_X) \longrightarrow \mathcal{O}_X(mK_X)$ *is surjective.*

If $\nu(X) = 0$, then the abundance conjecture is true ([Ka6]). Thus the conjecture is reduced to the following statement : X being a minimal variety, $\nu(X) > 0$ implies $\varkappa(X) > 0$. Miyaoka [Mi1,2] proved that $\varkappa(X) \geqq 0$ for a minimal algebraic variety X with dim X = 3. See also [W2].

The minimal model conjecture and the abundance conjecture combined imply the two basic conjectures in the classification theory of algebraic varieties, i.e., the Iitaka conjecture on the Kodaira dimension of algebraic fiber spaces ([Ka6]) and the invariance of plurigenera under deformations ([N]) (cf. [L]).

The minimal model conjectures are solved in some special cases.

THEOREM 5 ([T], [S3]). *The minimal model conjecture is true, if* $\dim X = 3$, $\dim S = 1$ *and if* $f : X \longrightarrow S$ *is a semi-stable degeneration of minimal surfaces* (cf. [Ku], [PP]).

THEOREM 6 ([S1], [KMM]). *The flip conjecture II is true in the case where* $\dim X = 3$ *or* 4.

If $\dim X = 3$, then this was proved by Shokurov [S1] ; the *difficulty* $d(X)$ drops after the flip, where $d(X)$ is the number of prime divisors on a desingularization of X whose discrepancies are less than 1. In case $\dim X = 4$, we use also the dimension $\rho_2(X)$ of the subspace of $H_4(X, \mathbb{Q})$ generated by the classes of algebraic 2-cycles. Then either $d(X') < d(X)$, *or* $d(X') = d(X)$ and $\rho_2(X') < \rho_2(X)$ ([KMM]).

Let X be a variety with at most canonical singularities *or* let (X, Δ) be a log-terminal pair, and consider a Weil divisor D on X. If the minimal or log-minimal model conjecture is true, respectively, then it follows easily that

$$\mathcal{R}(D) = \bigoplus_{m \geqq 0} \mathcal{O}_X(mD)$$

is finitely generated as a graded $\mathcal{O}_X$-algebra.

THEOREM 7 ([Ka8]). *Let* X *be a projective variety of dimension 3 such that* $(X, 0)$ *is log-terminal and let* D *be a Weil divisor on* X. *Then* $\mathcal{R}(D)$ *is finitely generated.*

In contrast to the two dimensional case, minimal models are not uniquely determined by their birational class.

THEOREM 8 ([Ka8]). *Let* X *and* X' *be two strictly terminal minimal algebraic varieties of dimension 3 and let* $\beta : X \dashrightarrow X'$ *be a birational map. Assume that* X *is good. Then* X' *is also good, and* β *can be realized as a composition of log-flips with respect to some* $\mathbb{Q}$*-divisor* Δ *on* X *and its strict transforms for which* K

is relatively numerically trivial.

THEOREM 9 ([Ka8]). *Let X be a terminal minimal algebraic variety of dimension 3 with $K_X \approx 0$. Then*

$$\chi(X, \mathcal{O}_X) = \sum_{P \in \mathrm{Sing}(X)} g(P) = 0,\ 1,\ 2,\ 3 \text{ or } 4,$$

where $g(P) = (r - 1/r)/24$ (resp. $= (r - 1/r)/12$) if P is an index r singular point whose canonical cover is non-singular (resp. a compound du Val singularity). Thus there is a constant m_0 such that $H^0(X, m_0K_X) \neq 0$ for all minimal algebraic varieties X of dimension 3 with $K_X \approx 0$.

Let $f : (X, \Delta) \longrightarrow S$ be a projective surjective morphism from a $\mathbb{Q}$-factorial log-terminal pair such that $K_X + \Delta$ is f-big. For each positive integer m with $m(K_X + \Delta) \in \mathrm{Div}(X)$, we define $\mathcal{O}(M_m)$ as the double dual of the image of the natural homomorphism $f^*f_*\mathcal{O}(m(K_X + \Delta)) \longrightarrow \mathcal{O}(m(K_X + \Delta))$, if it is not zero. Thus $m(K_X + \Delta) = M_m + F_m$ for some effective divisors F_m. We set

$$M = \lim M_m/m \quad \text{and} \quad F = \lim F_m/m \quad \text{in } \mathrm{Div}(X)_{\mathbb{R}}.$$

If M is f-nef, i.e., if the linear functional corresponding to M is non-negative on $\overline{NE}(X/S) - \{0\}$, then the expression $K_X + \Delta = M + F$ is called the *Zariski decomposition* (cf. [Z], [Ft1], [C], [Mw], [Ka7]).

THEOREM 10 ([Ka7]). *Let $f : (X, \Delta) \longrightarrow S$ be as above and assume the existence of the Zariski decomposition $K_X + \Delta = M + F$. Then*

$$\mathcal{R}(X/S, K_X + \Delta) = \bigoplus_{m \geq 0} f_*\mathcal{O}_X([m(K_X + \Delta)])$$

is finitely generated.

REFERENCES

[A] M. Artin, Some numerical criteria for contractibility of curves on algebraic surfaces, Amer. J. Math. 84(1962), 485-496.

[B] X. Benveniste, Sur l'anneau canonique de certaines variétés de dimension 3, Invent. Math. 73(1983), 157-164.

[Bo] E. Bombieri, Canonical models of surfaces of general type, Publ. Math. IHES 42(1973), 171-219.

[C] S. D. Cutkosky, Zariski decomposition of divisors on algebraic varieties, preprint, Brandeis Univ. 1985.

[E] R. Elkik, Rationalité des singularités canoniques, Invent. Math. 64(1981), 1-6.

[F] P. Francia, Some remarks on minimal models I, Compositio Math. 40(1980), 301-313.

[Fk] A. Fujiki, On the minimal models of complex manifolds, Math. Ann. 253(1980), 111-128.

[Ft1] T. Fujita, Zariski decomposition and canonical rings of elliptic threefolds, preprint, Tokyo Univ. 1984.

[Ft2] ————— , A relative version of Kawamata-Viehweg's vanishing theorem, preprint, Tokyo Univ. 1985.

[H] H. Hironaka, Resolution of singularities of an algebraic variety over a field of characteristic zero, Ann. of Math. 79(1964), 109-326.

[I] S. Iitaka, On D-dimension of algebraic varieties, J. Soc. Math. Japan 23(1971), 356-373.

[Ka1] Y. Kawamata, A generalization of Kodaira-Ramanujam's vanishing theorem, Math. Ann. 261(1982), 43-46.

[Ka2] ————— , On the finiteness of generators of a pluri-canonical ring for a 3-fold of general type, Amer. J. Math. 106(1984), 1503-1512.

[Ka3] ————— , Elementary contractions of algebraic 3-folds, Ann. of Math. 119(1984), 95-110.

[Ka4] ————— , The cone of curves of algebraic varieties, Ann. of Math. 119(1984), 603-633.

[Ka5] ————— , Pluricanonical systems on minimal algebraic varieties, Invent. Math. 79(1985), 567-588.

[Ka6] ————— , Minimal models and the Kodaira dimension of algebraic varieties, preprint, Tokyo Univ. 1985.

[Ka7] ————— , The Zariski decomposition of log-canonical divisors, preprint, Tokyo Univ. 1985.

[Ka8] ————— , Some properties of minimal algebraic 3-folds, in preparation.

[KMM] Y. Kawamata, K. Matsuki and K. Matsuda, Introduction to the minimal model problem, in preparation.

[Kl] S. Kleiman, Toward a numerical theory of ampleness, Ann. of Math. 84(1966), 293-344.

[K] K. Kodaira, Pluricanonical systems on algebraic surfaces of general type, J. Math. Soc. Japan 30(1968), 170-192.

[Ko] J. Kollár, The cone theorem : Note to [Ka4], Ann. of Math. 120(1984), 1-5.

[Ku] V. Kulikov, Degenerations of K3 surfaces and Enriques

surfaces, Math. USSR Izv. 11(1977), 957-989.

[L] M. Levine, Pluri-canonical divisors on Kähler manifolds, Invent. Math. 74(1983), 293-303.

[Mi1] Y. Miyaoka, Deformations of a morphism along a foliation and applications, preprint, Tokyo Metropolitan Univ. 1985.

[Mi2] ———, The pseudo-effectivity of $3c_2 - c_1^2$ for varieties with numerically effective canonical classes and the non-negativity of the Kodaira dimension of minimal threefolds, preprint, Tokyo Metropolitan Univ. 1985.

[MM] Y. Miyaoka and S. Mori, A numerical criterion of uniruledness, preprint.

[Mo] S. Mori, Threefolds whose canonical bundles are not numerically effective, Ann. of Math. 116(1982), 133-176.

[Mw] A. Moriwaki, Semi-ampleness of the numerically effective part of Zariski decomposition, preprint, Kyoto Univ. 1985.

[Mf] D. Mumford, The canonical ring of an algebraic surface, (Appendix to [Z]), Ann of Math. 76(1962), 612-615.

[N] N. Nakayama, Invariance of the plurigenera of algebraic varieties under minimal model conjectures, preprint, Tokyo Univ. 1984.

[PP] U. Persson and H. Pinkham, Degeneration of surfaces with trivial canonical bundle, Ann. of Math. 113(1981), 45-66.

[Re1] M. Reid, Canonical 3-folds, in Géométrie Algébriques Angers 1979, A. Beauville ed., 1980, Sijthoff & Noordhoff, Alphen aan den Rijn, The Netherlands, 273-310.

[Re2] ———, Minimal models of canonical 3-folds, in Algebraic Varieties and Analytic Varieties, S. Iitaka ed., Advanced Studies in Pure Math. 1(1983), Kinokuniya, Tokyo, and North-Holland, Amsterdam, 131-180.

[Re3] ———, Projective morphisms according to Kawamata, preprint, Warwick Univ. 1983.

[S1] V. V. Shokurov, Theorem on non-vanishing, Math. USSR Izv. 19(1985), ?

[S2] ———, On the closed cone of curves of algebraic 3-folds, Math. USSR Izv. 24(1985), 193-198.

[S3] ———, Letter to M. Reid, May 24, 1985.

[T] S. Tsunoda, Projective degenerations of algebraic surfaces, in preparation.

[U] K. Ueno, Classification Theory of Algebraic Varieties and Compact Complex Spaces, Lecture Notes in Math. 439(1975), Springer, Berlin-Heidelberg-New York.

[V] E. Viehweg, Vanishing theorems, J. reine angew. Math. 335(1982), 1-8.

[W1] P. M. H. Wilson, On the canonical ring of algebraic varieties, Compositio Math. 43(1981), 365-385.

[W2] ———, On regular threefolds with $\varkappa = 0$, Invent. Math. 76(1984), 345-355.

[Z] O. Zariski, The theorem of Riemann-Roch for high multiples of an effective divisor on an algebraic surface, Ann. of Math. 76(1962), 560-615.

Modulräume holomorpher Abbildungen auf komplexen Mannigfaltigkeiten mit 1-konkavem Rand

von

Siegmund Kosarew

Einleitung. In der Arbeit $[Ko]_1$ wurde ein Modulraum für holomorphe S-Abbildungen $f : (X,A) \longrightarrow Y$ konstruiert, wobei $X \xrightarrow{p} S$ und $Y \xrightarrow{q} S$ holomorphe Abbildungen in einen komplexen Raumkeim $S = (S,0)$ sind, p flach und $A \subset X(0) = p^{-1}(0)$ eine kompakte konkave Teilmenge ist. Hierbei mußte der Konkavitätsgrad einer gewissen Abschätzung im Vergleich zur Tiefe von $\mathcal{O}_{X(0)}$ genügen. Das Ziel dieser Arbeit ist es nun eine entsprechende Aussage für den allgemeineren Konkavitätsbegriff von Commichau-Grauert (vgl. [Co-Gr] § 2), aber für speziellere, nämlich glatte Abbildungen p und q zu beweisen. Ist φ eine definierende Funktion des Randes von A in Umgebung von $x \in \partial A$, so fordert man, daß die Leviform von φ in x, eingeschränkt auf den holomorphen Tangentialraum von φ, mindestens einen negativen Eigenwert besitzt. Die in $[Ko]_1$ angewandte Beweismethode überträgt sich leider nicht auf diesen Fall, da man die dortige Aussage (7.6) hier nicht zeigen kann. Es werden statt dessen Hilbert-Raum Techniken verwandt (auch die Lösung des $\bar{\partial}$ Problems nach Hörmander), die es nach Überwindung einiger technischer Probleme erlauben, die gewünschte Spaltungsaussage zu zeigen. Die typischen banachanalytischen Schlüsse und Techniken finden hier kaum Verwendung und werden im wesentlichen nur an einer Stelle in Abschnitt 5 benutzt. Im letzten Abschnitt 6 geben wir noch eine Vergleichsaussage für 1-positive Vektorraumbündel in Umgebung des Nullschnitts an, die in $[Ko]_2$ benutzt wird.

Komplexe Räume werden stets als hausdorffsch und mit abzählbarer Topologie vorausgesetzt.

1. Formulierung der Hauptsätze.

(1.1) Satz. Seien $S = (S,0)$ ein komplexer Raumkeim, $p : X \longrightarrow S$ und $q : Y \longrightarrow S$ glatte holomorphe Abbildungen sowie $A \subset X(0) = p^{-1}(0)$ ein Kompaktum mit 1-konkavem glatten Rand (vgl. (2.1)). Ferner sei $f_o : (X(0),A) \longrightarrow Y(0)$ eine holomorphe Abbildung.

Dann ist der mengenwertige Funktor

$$T \longmapsto \mathrm{Hom}_T((X_T,A),Y_T)^o$$

aller holomorphen T-Abbildungen $f : (X_T,A) \longrightarrow Y_T$ mit $f(0) = f_o$ und $X_T = T \times_S X$, $Y_T = T \times_S Y$ auf der Kategorie der komplexen Raumkeime über S darstellbar.

Es gilt auch die folgende globale Version

(1.2) Satz. Seien S ein komplexer Raum und $p : X \longrightarrow S$, $q : Y \longrightarrow S$ glatte holomorphe Abbildungen komplexer Räume sowie $A \subset X$ eine eigentlich über S liegende abgeschlossene Teilmenge von X, so daß jede Faser $A(s)$, $s \in S$, 1-konkaven glatten Rand besitzt.*)

Dann ist der mengenwertige Funktor

$$T \longmapsto \mathrm{Hom}_T((X_T,A_T),Y_T)$$

auf der Kategorie (An/S) der komplexen Räume über S darstellbar.

Wir werden zunächst den Satz (1.1) beweisen und hieraus mit Hilfe eines Darstellbarkeitskriteriums von Schuster/Vogt [S-V], Proposition (1.1), den Satz (1.2) folgern.

2. Relativ kompakte Teilmengen mit 1-konkavem Rand

(2.1) Sei X eine n-dimensionale komplexe Mannigfaltigkeit und $B \subset\subset X$ eine relativ kompakte offene Teilmenge mit glattem $\mathcal{C}^3$ Rand

*) Hierbei soll zusätzlich die Abhängigkeit einer definierenden Funktion des Randes von $A(s)$ differenzierbar in s sein.

∂B. Wir nennen ∂B 1-<u>konkav</u>, falls folgendes gilt:

Sei $x_o \in \partial B$ und $U \subset X$ eine offene Umgebung von x_o sowie $\varphi \in \Gamma(U, \mathcal{E}_X)$ eine reellwertige differenzierbare Funktion mit $(D\varphi)(x_o) \neq 0$ und $B \cap U = \{x \in U : \varphi(x) < 0\}$. Dann besitzt die Leviform $L(\varphi)$ in x_o, eingeschränkt auf den komplex (n-1)-dimensionalen holomorphen Tangentialraum $T_{\varphi,x_o} \subset T_{x_o}(X)$ von φ in x_o mindestens einen negativen Eigenwert.

<u>(2.2) Bemerkung.</u> (1) Nach Verkleinerung von U als Umgebung von x_o kann man annehmen, daß die Leviform $L(\varphi)$ auf $T_{\varphi,x}$ mindestens einen negativen Eigenwert besitzt für alle Punkte $x \in \partial B \cap U$.

(2) Es gibt eine offene Umgebung U von $\overline{B}$ in X und eine reellwertige differenzierbare Funktion $\varphi \in \Gamma(U, \mathcal{E}_X)$ mit $(D\varphi)(x) \neq 0$ für alle $x \in \partial B$ und $B = \{x \in U : \varphi(x) < 0\}$ sowie der Eigenschaft (1). Ferner gilt $A := \overline{B} = \{x \in U : \varphi(x) \leq 0\}$.

Zum <u>Beweis</u> sei auf [Gu-Ro] chap. IX, A oder [An-Gr] Proposition 15 verwiesen.

Die folgende Aussage beschreibt das lokale Fortsetzungsverhalten holomorpher Funktionen in der Nähe des Randes von B.

<u>(2.3) Proposition.</u> Sei X eine n-dimensionale komplexe Mannigfaltigkeit mit $n \geq 2$ und $B \subset\subset X$ eine relativ kompakte Teilmenge mit glattem 1-konkavem Rand. Zu jedem Punkt $x_o \in \partial B$ gibt es dann offene Umgebungen P, U mit $P \subset\subset U$ und eine biholomorphe Abbildung $\Phi : U \longrightarrow V \subset\subset \mathbb{C}^n$ mit $\Phi(x_o) = 0$ sowie einen offenen Polyzylinder $D = D' \times D'' \subset \mathbb{C} \times \mathbb{C}^{n-1}$ um 0 in V mit $\Phi(P) = D$, so daß gilt

(1) Es gibt eine konzentrische Schrumpfung D_1' von D' und eine Schrumpfung D_1'' von D'' mit $0 \notin D_1''$ derart, daß der Abschluß der Menge

$$\tilde{D} := (D' \times D_1'') \cup [(D' \setminus \overline{D}_1') \times D'']$$

in $\Phi(\overline{B} \cap U)$ enthalten ist.

(2) Die Beschränkungsabbildung

$$\Gamma(P, \mathcal{O}_X) \longrightarrow \Gamma(B \cap P, \mathcal{O}_X)$$

ist bijektiv.

Beweis. Wir dürfen annehmen, daß X eine offene Nullumgebung im $\mathbb{C}^n$ ist mit $x_o = 0$. Die Taylorentwicklung von φ um 0 bis zur zweiten Ordnung lautet dann

$$\varphi(z) = 2\,\mathrm{Re}\Big[\sum_{i=1}^{n} \frac{\partial\varphi}{\partial z_i}(0)\, z_i + \sum_{i,j=1}^{n} \frac{\partial^2\varphi}{\partial z_i \partial z_j}(0)\, z_i z_j + \sum_{i,j=1}^{n} \frac{1}{2}\frac{\partial^2\varphi}{\partial z_i \partial \bar z_j}(0)\, z_i \bar z_j\Big] + r(z)$$

mit $\lim_{z \to 0} (r(z)/|z|^2) = 0$. Wegen der Voraussetzung gibt es ein ξ aus $T_{\varphi,o}$ mit $|\xi| = 1$ und

$$\sum_{i,j=1}^{n} \frac{\partial^2\varphi}{\partial z_i \partial \bar z_j}(0)\, \xi_i \bar\xi_j < -K$$

für ein $K > 0$. Wir dürfen ohne Einschränkung $\frac{\partial\varphi}{\partial z_1}(0) \neq 0$ annehmen. Wir schreiben ξ in der Form $\xi = (\xi_1, \xi")$ und betrachten die Kurve $C \subset \mathbb{C}^n$ mit der Parameterdarstellung $f(t) := (\xi_1 t + at^2, t\xi")$ für $t \in \mathbb{C}$, wobei die Konstante a noch festgelegt wird. Für t nahe bei 0 gilt dann

$$\varphi(f(t)) = |t|^2 \sum_{i,j=1}^{n} \frac{\partial^2\varphi}{\partial z_i \partial \bar z_j}(0)\, \xi_i \bar\xi_j + 2\,\mathrm{Re}\Big[t^2 \Big(\frac{\partial\varphi}{\partial z_1}(0)\, a + \sum_{i,j=1}^{n} \frac{\partial^2\varphi}{\partial z_i \partial z_j}(0)\xi_i \xi_j\Big)\Big] + o(|t|^2)\,.$$

Setzen wir nun $a := -\Big(\frac{\partial\varphi}{\partial z_1}(0)\Big)^{-1} \sum_{i,j=1}^{n} \frac{\partial^2\varphi}{\partial z_i \partial z_j}(0)\, \xi_i \xi_j$, so folgt

$$\varphi(f(t)) \le - K'|t|^2$$

für eine Konstante $K' > 0$ und alle $t \in \mathbb{C}$ mit $|t| < \varepsilon$. Es gibt damit also ein glattes Kurvenstück C_1 in $\overline{B}$ durch 0 mit $C_1 \setminus \{0\} \subset B$.

Nach einer holomorphen Koordinatentransformation kann man erreichen, daß C_1 die z_1-Achse ist. Es gilt dann für alle z_1 nahe 0

$$\varphi(z_1,0) = \begin{cases} 0, & \text{für } z_1 = 0 \\ <0, & \text{für } z_1 \neq 0 . \end{cases}$$

Insbesondere ist $\frac{\partial\varphi}{\partial z_1}(0) = 0$. Nach Umnummerieren der Koordinaten kann man $\frac{\partial\varphi}{\partial z_2}(0) \neq 0$ annehmen. Ferner gibt es Zahlen $r_1, s_1, r'' > 0$ mit $r_1 > s_1$, so daß $\varphi(z_1,z'') < 0$ ist für alle (z_1,z'') mit $s_1 \le |z_1| \le r_1$ und $|z''| \le r''$. Sei $\eta'' := (\eta_2,0)$ ein Punkt. Dann ist

$$\varphi(z_1,\eta'') = \varphi(z_1,0) + a(z_1,\eta_2) + O(|\eta_2|^2)$$

mit $a(z_1,\eta_2) = 2\,\mathrm{Re}\left[\frac{\partial\varphi}{\partial z_2}(z_1,0)\,\eta_2\right]$. Nach eventuellem Verkleinern von r_1, s_1 und r'' kann man annehmen, daß $a(z_1,t\eta_2) < 0$ ist für alle $t \in \,]0,1]$ und alle z_1 mit $|z_1| \le r_1$, für geeignete Wahl von η_2. Es existiert dann ein $t_o \in \,]0,1]$ mit $\varphi(z_1,t_o\eta'') < 0$ für alle z_1 mit $|z_1| \le r_1$ und $|t_o\eta_2| < r''$. Wir setzen nun $D' := \{z_1 \in \mathbb{C} : |z_1| < r_1\}$, $D_1' := \{z_1 \in \mathbb{C} : |z_1| < s_1\}$ und $D'' := \{z'' \in \mathbb{C}^{n-1} : |z''| < r''\}$ sowie $D_1'' := t_o\eta'' + E_\tau$, wobei E_τ der Polyzylinder vom Radius τ um Null in $\mathbb{C}^{n-1}$ ist. Für genügend kleines τ ist dann die Eigenschaft (1) der Behauptung erfüllt.

Die Fortsetzungsaussage (2) folgt sofort aus der Tatsache, daß die Beschränkungsabbildung $\Gamma(D, \mathcal{O}_{\mathbb{C}^n}) \longrightarrow \Gamma(\tilde{D}, \mathcal{O}_{\mathbb{C}^n})$ bijektiv ist (und $B \cap D$ als zusammenhängend angenommen werden darf). -

<u>(2.4) Bemerkung.</u> (1) Die Behauptungen in (2.3) bleiben gültig, falls man bei festgehaltenem Φ kleine $\mathcal{C}^2$ - Störungen von $\varphi|U$ und kleine Veränderungen der Radien von D', D'', D_1' und D_1'' zuläßt.

(2) In der Situation von (2.3) ist die Beschränkungsabbildung

$$\Gamma_{L^2}(P, \mathcal{O}_X) \longrightarrow \Gamma_{L^2}(B \cap P, \mathcal{O}_X)$$

zwischen den quadratintegrierbaren holomorphen Funktionen bijektiv.

Beweis. Der Teil (1) folgt aus dem Beweis von (2.3). Zu (2). Es genügt die Surjektivität der Beschränkungsabbildung auf den quadratintegrierbaren holomorphen Funktionen einzusehen. Wir dürfen dabei ohne Einschränkung $D = P$ und $B \cap P = \tilde{D}$ annehmen. Sei also f aus $\Gamma(D, \mathcal{O}_{\mathbb{C}^n})$ eine holomorphe Funktion mit der (konvergenten) Reihenentwicklung $f(z_1, z'') = \sum_{\nu \in \mathbb{N}} f_\nu(z'') z_1^\nu$. Wir legen dabei die im Beweis von (2.3) eingeführten Bezeichnungsweisen zugrunde. Ist $f|\tilde{D}$ und damit insbesondere $f|(D' \setminus \overline{D}_1') \times D''$ quadratintegrierbar, so folgt $\|f\|^2_{(D' \setminus \overline{D}_1') \times D''} = \sum_{\nu \in \mathbb{N}} \|f_\nu\|^2_{D''} \|z_1^\nu\|^2_{D' \setminus \overline{D}_1'} < \infty$. Es gibt nun eine Konstante $C > 0$ mit $\|z_1^\nu\|_{D'} \leq C \|z_1^\nu\|_{D' \setminus \overline{D}_1'}$ für alle $\nu \in \mathbb{N}$. Damit gilt aber $\|f\|_D < \infty$.

(2.5) Seien X und B wie in (2.3). Unter einem Meßatlas für B verstehen wir eine endliche Familie $\Phi_i : U_i \longrightarrow V_i \subset\subset \mathbb{C}^n$, $i \in I$, von Karten von X zusammen mit offenen Teilmengen $P_i \subset\subset U_i$, so daß folgende Bedingungen erfüllt sind

(1) $D_i := \Phi_i(P_i)$ ist ein offener Polyzylinder im $\mathbb{C}^n$.

(2) $\overline{B} \subset \bigcup_{i \in I} P_i$ und $\overline{B} \cap P_i \neq \emptyset$ für alle $i \in I$.

(3) Ist $\partial B \cap P_i \neq \emptyset$, so erfüllen die Daten (Φ_i, P_i, D_i) die Eigenschaften in Proposition (2.3).

Wir setzen überdies $Q_i := B \cap P_i$ und $C_i := \Phi_i(Q_i)$ für $i \in I$. Das so definierte System von Daten kürzen wir mit $\underline{Q}$ ab.

Ist ferner eine lokalfreie Garbe $\mathcal{L}$ vom Rang g in Umgebung von $A := \overline{B}$ gegeben, so nennen wir $\underline{Q}$ angepaßt an $\mathcal{L}$, wenn zusätzlich Trivialisierungen $\mathcal{L}|U_i \xrightarrow{\sim} \mathcal{O}^g_{U_i}$ fixiert sind für alle $i \in I$.

Unter einer Schrumpfung $\underline{Q}'$ des Meßatlas $\underline{Q}$ verstehen wir folgendes: Gegeben sind konzentrische Schrumpfungen D_i' von D_i und ein $c < 0$, so daß man mit den neuen Daten $P_i' := \Phi_i^{-1}(D_i')$, $B' = B_c :=$ $\{x \in B : \varphi(x) < c\}$ und $Q_i' := B_c \cap P_i'$ wieder einen Meßatlas $\underline{Q}'$ in dem eben definierten Sinne erhält. Offenbar besitzt B beliebig feine Meßatlanten und jeder Meßatlas von B ist schrumpfbar.

3. L^2 - Kohomologie

(3.1) Sei $U \subset\subset \mathbb{C}^n$ eine offene beschränkte Teilmenge. Für eine holomorphe Funktion $f \in \Gamma(U, \mathcal{O}_{\mathbb{C}^n})$ sei

$$\|f\|_U^2 := \int_U f(z)\overline{f}(z)\, d\mu$$

die Hilbertpseudonorm auf $\Gamma(U, \mathcal{O}_{\mathbb{C}^n})$, wobei μ das Lebesgue-Maß auf dem $\mathbb{C}^n$ bezeichnet. Wir fixieren ein $k > n$ und setzen

$$|||f|||_U^2 := \sum_{\substack{\alpha \in \mathbb{N}^n \\ |\alpha| \leq k}} \|D^\alpha f\|_U^2 .$$

Offenbar ist $|||.|||_U$ wiederum eine Hilbertpseudonorm auf $\Gamma(U, \mathcal{O}_{\mathbb{C}^n})$ und wir bezeichnen mit $\Gamma_{L^2}(U, \mathcal{O}_{\mathbb{C}^n})$ bzw. $\Gamma_h(U, \mathcal{O}_{\mathbb{C}^n})$ den Hilbertraum der bzgl. $\|.\|_U$ bzw. $|||.|||_U$ beschränkten Elemente von $\Gamma(U, \mathcal{O}_{\mathbb{C}^n})$. Die Inklusion $\Gamma_h(U, \mathcal{O}_{\mathbb{C}^n}) \longrightarrow \Gamma_{L^2}(U, \mathcal{O}_{\mathbb{C}^n})$ ist stetig. Erfüllt U die sogenannte Kegeleigenschaft aus [Ad] p. 66 (z.B. U ein Polyzylinder), so gibt es eine Konstante $K > 0$ mit $|||fg|||_U \leq K\, |||f|||_U\, |||g|||_U$ für alle f,g aus $\Gamma_h(U, \mathcal{O}_{\mathbb{C}^n})$, vgl. loc. cit. p. 115.

(3.2) Gegeben sei nun eine offene Teilmenge $B \subset\subset X$ einer n-dimensionalen komplexen Mannigfaltigkeit X mit 1-konkavem glatten Rand ∂B. Ferner sei $\underline{Q}$ ein Meßatlas von B mit Karten $\Phi_i : U_i \longrightarrow V_i$, $i \in I$, und $\mathcal{L}$ eine lokalfreie Modulgarbe in Umgebung von $\overline{B}$. Wir setzen voraus, daß $\underline{Q}$ angepaßt an $\mathcal{L}$ ist. Es bezeichne dann $(C_h^\bullet(\underline{Q},\mathcal{L}),\delta)$ bzw. $(C_{L^2}^\bullet(\underline{Q},\mathcal{L}),\delta)$ den Čech-Komplex der bzgl. $|||\,.\,|||$ bzw.

$\|.\|$ quadratintegrierbaren Koketten von $\mathcal{L}$ auf der offenen Überdeckung $(Q_i)_{i\in I}$ von B. Hierbei ist die jeweilige Hilbertpseudonorm auf $\Gamma(Q_{i_0} \cap \ldots \cap Q_{i_r}, \mathcal{L})$ bzgl. der Karte Φ_{i_0} und der Trivialisierung $\mathcal{L}|Q_{i_0} \cong \mathcal{O}^{rg(\mathcal{L})}_{Q_{i_0}}$ zu nehmen.

(3.3) Satz. In der Situation von (3.2) ist die Kohomologiegruppe $H^r_h(\underline{Q}, \mathcal{L})$ endlichdimensional für $r = 0$ und separiert für $r = 1$.

Beweis. Sei φ eine definierende Funktion für B wie in (2.2)(2), d.h. insbesondere ist $B = \{x \in W : \varphi(x) < 0\}$ für eine offene Umgebung W von $\overline{B}$ in X, und $c < 0$ so klein, so daß die durch $\tilde{B} := \{x \in W : \varphi(x) < c\}$ und $\tilde{Q}_i := \tilde{B} \cap P_i$, $i \in I$, definierten Daten einen Meßatlas für $\tilde{B}$ liefern. Aus (2.3) folgt die Bijektivität der Beschränkungsabbildung $\Gamma(B, \mathcal{L}) \longrightarrow \Gamma(\tilde{B}, \mathcal{L})$ und daher gilt $\dim_{\mathbb{C}} \Gamma(B, \mathcal{L}) < \infty$ nach einem bekannten Kompaktheitsargument. Offensichtlich ist $H^0_h(\underline{Q}, \mathcal{L}) = \Gamma(B, \mathcal{L})$.

Zum Nachweis der Separiertheit von $H^1_h(\underline{Q}, \mathcal{L})$ betrachten wir die Sequenz $H^1_h(\underline{Q}, \mathcal{L}) \to H^1_{L^2}(\underline{Q}, \mathcal{L}) \to H^1(\underline{Q}, \mathcal{L})$ von topologischen Vektorräumen. Es genügt folgendes zu zeigen

(a) $H^1_h(\underline{Q}, \mathcal{L}) \longrightarrow H^1(\underline{Q}, \mathcal{L})$ ist injektiv,

(b) $H^1_{L^2}(\underline{Q}, \mathcal{L})$ ist separiert.

Zu (a). Wir fixieren eine Schrumpfung $\underline{Q}'$ von $\underline{Q}$ gemäß (2.5). Sei $\xi = (\xi_{ij})$ aus $Z^1_h(\underline{Q}, \mathcal{L})$ gegeben mit $\xi = \delta\eta$ und η aus $C^0(\underline{Q}, \mathcal{L})$. Wir zeigen $|||\eta_i|||_{Q_i} < \infty$. Wegen $Q'_i \subset\subset Q_i$ gilt $|||\eta_i|||_{Q'_i} < \infty$. Ist $j \in I$ beliebig, so hat man $\eta_i = \eta_j - \xi_{ij}$ auf $Q_i \cap Q_j$ und damit $|||\eta_i|||_{Q_i \cap Q'_j} \leq K\, |||\eta_j|||_{Q'_j} + |||\xi|||$, wobei K eine Konstante ist, die nur von dem definierenden Kozyklus von $\mathcal{L}$ auf $P_i \cap P_j$ abhängt. Da nun $Q_i \cap B' = \bigcup_{i\in I} Q_i \cap Q'_j$ gilt, folgt

$|||\eta_i|||_{Q_i \cap B'} < \infty$, nach (2.4) (2) dann aber auch $|||\eta_i|||_{Q_i} < \infty$.

Zu (b). Die Separiertheit von $H^1_{L^2}(\underline{Q},\mathcal{L})$ ist gleichbedeutend mit der Abgeschlossenheit von $B^1_{L^2}(\underline{Q},\mathcal{L})$ in $C^1_{L^2}(\underline{Q},\mathcal{L})$. Seien $\sigma_i : X \to \mathbb{R}_{\geq 0}$, $i \in I$, differenzierbare Funktionen mit Träger $(\sigma_i) \subset\subset P_i$ und $\sum_{i\in I} \sigma_i = 1$ in Umgebung von $\overline{B}$. Wir nehmen ein $\xi = (\xi_{ij})$ aus $B^1_{L^2}(\underline{Q},\mathcal{L})$ und ein η aus $C^0_{L^2}(\underline{Q},\mathcal{L})$ mit $\xi = \delta\eta$. Setzen wir $\alpha_i := \sum_{k\in I} \sigma_k \xi_{ki}$, so ist α_i ein differenzierbarer Schnitt von $\mathcal{L}$ über Q_i und für $\alpha = (\alpha_i)_{i\in I}$ gilt $\delta\alpha = \xi$. Sei $z_i := \bar{\partial}\alpha_i$. Wegen $z_i|Q_i \cap Q_j = z_j|Q_i \cap Q_j$ verheften sich die z_i zu einem Element z aus $Z^1_{\bar{\partial}}(\Gamma_{L^2}(B,\mathcal{E}^{0,\bullet}_X(\mathcal{L})))$ mit $\|z\| \leq C\|\xi\|$ mit einer von z unabhängigen Konstante $C > 0$. Da nun z im Bild des dicht definierten Operators $\bar{\partial} : \Gamma_{L^2}(B,\mathcal{E}^{0,0}(\mathcal{L})) \to Z^1_{\bar{\partial}}(\Gamma_{L^2}(B,\mathcal{E}^{0,\bullet}_X(\mathcal{L})))$ liegt, gibt es nach $[\text{Hö}]_1$ Beweis von Theorem 3.4.1 (mit $q = 0$) sowie 1.1.2 und 1.1.3 ein w aus $\Gamma_{L^2}(B,\mathcal{E}^{0,0}_X(\mathcal{L}))$ mit $\bar{\partial}w = z$ und $\|w\| \leq C'\|z\|$ mit einer von z unabhängigen Konstanten $C' > 0$ (vgl. auch $[\text{Hö}]_2$ Beweis von Lemma 4.1.1, p. 78). Sei nun $\gamma_i := \alpha_i - w|Q_i$ und $\gamma := (\gamma_i)_{i\in I}$. Dann folgt $\delta\gamma = \delta\alpha = \xi$ und $\bar{\partial}\gamma_i = \bar{\partial}\alpha_i - \bar{\partial}w|Q_i = \bar{\partial}\alpha_i - z_i = 0$ sowie $\|\gamma\| \leq \|\alpha\| + \|w\| \leq C''\|\xi\|$, wobei C'' nur von C, C' und den σ_i abhängt. Hieraus folgt die Abgeschlossenheit des Bildes von $\delta : C^0_{L^2}(\underline{Q},\mathcal{L}) \to C^1_{L^2}(\underline{Q},\mathcal{L})$ und (3.3) ist damit bewiesen.

Die folgende Aussage wird später in Abschnitt 5 entscheidend benutzt werden.

<u>(3.4) Satz.</u> Sei $\underline{Q}$ ein Meßatlas von $B \subset X$, der angepaßt an $\mathcal{L}$ ist, und $\underline{Q}'$ eine Schrumpfung von $\underline{Q}$. Dann ist der Čech-Korandoperator

$$\hat{\delta}: \prod_{i\in I} \Gamma_h(Q_i,\mathcal{L}) \longrightarrow \prod_{(i,j)\in I^2} \Gamma_h(Q_i \cap Q'_j,\mathcal{L})$$

ein direkter Homomorphismus von Hilbert-Räumen und Kern$(\hat{\delta})$ ist endlichdimensional.

<u>Beweis.</u> Es gilt Kern$(\hat{\delta}) = H^0_h(\underline{Q},\mathcal{L}) = \Gamma(B,\mathcal{L})$ und $\dim_{\mathbb{C}} H^0_h(\underline{Q},\mathcal{L}) < \infty$ nach (3.3). Somit ist noch die Abgeschlossenheit von Bild $(\hat{\delta})$

zu zeigen. Wir setzen abkürzend $E := \prod_{(i,j)\in I^2} \Gamma_h(Q_i \cap Q'_j, \mathcal{L})$ und $Z(E) := E \cap \operatorname{Kern}(\delta)$. Wir können $\hat{\delta}$ als die Komposition

$$C_h^o(\underline{Q},\mathcal{L}) \xrightarrow{\beta^o} C_h^o(\underline{Q} \cap B',\mathcal{L}) \xrightarrow{\delta} Z_h^1(\underline{Q} \cap B',\mathcal{L}) \xrightarrow{\beta^1} Z(E)$$

schreiben, wobei β^o und β^1 Beschränkungsabbildungen sind. Nach (2.4) ist β^o bijektiv. Es genügt dann wegen (3.3) die Bijektivität von β^1 einzusehen. Die Injektivität ist hierbei evident. Zur Surjektivität: Sei $\xi = (\xi_{ij})$ aus $Z(E)$ vorgegeben, d.h. es gilt

$$\xi_{ik} = \xi_{ij} + \xi_{jk} \quad \text{auf} \quad Q_i \cap Q'_j \cap Q'_k .$$

Wir setzen zunächst jedes ξ_{ij} holomorph nach $Q_i \cap Q_j \cap B'$ fort: Es ist $B' = \bigcup_{k\in I} Q'_k$. Daher wird durch

$$\text{(3.4.1)} \qquad \xi_{ij}^{(k)} := \xi_{ik} - \xi_{jk}$$

ein holomorpher Schnitt von $\mathcal{L}$ auf $Q_i \cap Q_j \cap Q'_k$ definiert. Wir zeigen $\xi_{ij}^{(k)} = \xi_{ij}^{(l)}$ auf $Q_i \cap Q_j \cap Q'_k \cap Q'_l$. Man hat $\xi_{il} = \xi_{ik} + \xi_{kl}$ auf $Q_i \cap Q'_k \cap Q'_l$ und $\xi_{jl} = \xi_{jk} + \xi_{kl}$ auf $Q_j \cap Q'_k \cap Q'_l$. Über $Q_i \cap Q_j \cap Q'_k \cap Q'_l$ gilt dann

$$\begin{aligned} \xi_{ij}^{(k)} &= \xi_{ik} - \xi_{jk} \\ &= (\xi_{il} - \xi_{kl}) - (\xi_{jl} - \xi_{kl}) \\ &= \xi_{il} - \xi_{jl} \\ &= \xi_{ij}^{(l)} . \end{aligned}$$

Bezeichnet $\tilde{\xi} = (\tilde{\xi}_{ij})$ die so erhaltene Fortsetzung von ξ auf $C^1(\underline{Q} \cap B',\mathcal{L})$, so gilt $\delta\tilde{\xi} = 0$, da man $\xi_{ik}^{(l)} = \xi_{ij}^{(l)} + \xi_{jk}^{(l)}$ auf $Q_i \cap Q_j \cap Q_k \cap Q'_l$ für jedes $i \in I$ hat, wie sich sofort aus der Definition (3.4.1) ergibt. Ferner zeigt (3.4.1), daß mit ξ auch die Fortsetzung $\tilde{\xi}$ quadratintegrierbar bzgl. $|\!|\!|.|\!|\!|$ ist.

4. Beschreibung holomorpher S-Abbildungen

(4.1) Sei $S = (S,0)$ ein komplexer Raumkeim, $p : X \longrightarrow S$ eine glatte holomorphe Abbildung mit n-dimensionaler ausgezeichneter

Faser $X(0)$ und $B \subset\subset X(0)$ eine offene Teilmenge mit glattem 1-konkaven Rand. Gegeben sei ferner eine endliche Familie $\Phi_i : U_i \longrightarrow S \times V_i$, $i \in I$, von S-Isomorphismen, wobei $U_i \subset X$ offen ist, zusammen mit offenen Polyedern $P_i \subset U_i(0)$ derart, daß $\varphi_i := \Phi_i(0) : U_i(0) \longrightarrow V_i$ mit P_i, $i \in I$, einen Meßatlas für B bilden. Es bezeichne $\Phi_{ji} : \Phi_i(U_i \cap U_j) \longrightarrow \Phi_j(U_i \cap U_j)$ den Kartenwechsel. Setzen wir $V_{ij} := \varphi_i(U_i(0) \cap U_j(0))$ und gehen wir zum Keim über, so ist $\Phi_{ji} : S \times V_{ij} \longrightarrow S \times V_{ji}$ ein S-Isomorphismus.

Sei nun $\underline{Q}'$ eine Schrumpfung von $\underline{Q}$ gemäß (2.5) und $C_{ij} := = \varphi_i(Q_i \cap Q_j)$ sowie $C_{i\underline{j}} := \varphi_i(Q_i \cap Q_j^!)$. Dann liegt $\varphi_{ji}(C_{i\underline{j}})$ relativ kompakt in C_j und für einen geeigneten Repräsentanten S' von S induziert Φ_{ji} eine holomorphe S-Abbildung

$$\Phi_{ji} : S' \times C_{i\underline{j}} \longrightarrow S' \times C_j \quad .$$

Wir sagen, daß eine Familie $(\Phi_i)_{i\in I}$ von Karten für p das <u>Polyederaxiom</u> erfüllt, falls aus $P_i \cap P_j \neq \emptyset$ die Inklusion $(P_i \cup P_j) \subset (U_i \cap U_j)$ folgt. Dies hat zur Konsequenz, daß der Kartenwechsel Φ_{ji} induziert ist durch eine holomorphe S-Abbildung $S \times D_i \longrightarrow S \times V_j$, die wir ebenfalls mit Φ_{ji} bezeichnen.

(4.2) Sei nun $q: Y \longrightarrow S$ eine weitere glatte holomorphe Abbildung mit m-dimensionaler ausgezeichneter Faser $Y(0)$ und $f_o : (X(0), A) \longrightarrow Y(0)$, mit $A := \overline{B}$, eine im folgenden feste holomorphe Abbildung. Gegeben seien außerdem Karten $(\Phi_i)_{i\in I}$ für $p : X \longrightarrow S$ wie in (4.1) und Karten $\Psi_\alpha : \tilde{U}_\alpha \longrightarrow S \times \tilde{V}_\alpha$, $\alpha \in \tilde{I}$, für q zusammen mit Polyedern $\tilde{P}_\alpha \subset\subset \tilde{U}_\alpha(0)$, so daß $\Psi_\alpha(0)(\tilde{P}_\alpha) =: \tilde{D}_\alpha$ ein offener beschränkter Polyzylinder im $\mathbb{C}^m$ ist. Wir setzen voraus, daß folgende Bedingungen erfüllt sind

(1) $(\Psi_\alpha)_{\alpha\in\tilde{I}}$ genügt dem Polyederaxiom.

(2) Es gibt eine Abbildung $\tau : I \longrightarrow \tilde{I}$ mit

$f_o(P_i) \subset\subset \tilde{P}_{\tau(i)}$ für alle $i \in I$.

Offenbar kann man zu gegebenem f_o stets solche Familien $(\Phi_i)_{i\in I}$ und $(\Psi_\alpha)_{\alpha\in\tilde{I}}$ von Karten von p bzw. q finden.

Wir setzen ferner $\mathcal{L} := f_o^* \Theta_{Y_o}$, wobei Θ_{Y_o} die Garbe der holomorphen Vektorfelder auf Y_o ist. Bzgl. dieser Daten ist

$(\varphi_i)_* (\mathcal{L}|U_i(0))$ kanonisch trivialisiert.

(4.3) Sei $T \to S$ ein komplexer Raumkeim über S und $f:(X_T, A) \to Y_T$ ein T-Morphismus mit $f(0) = f_o$. Verwenden wir die in (4.2) eingeführten Karten, so haben wir für jedes $i \in I$ ein kommutatives Diagramm

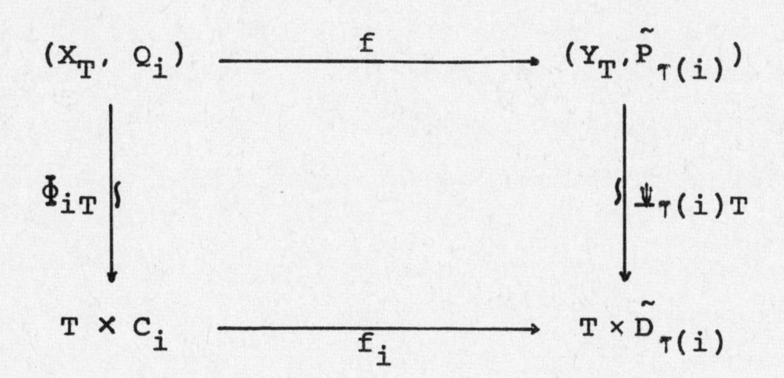

und es ist $f_i(0) = (f_o)_i$. Also gilt

$$f_j \Phi_{jiT} = \Psi_{\tau(j)\tau(i)T} f_i \quad \text{auf} \quad T \times C_{ij},$$

(4.3.1)

$$f_i(0) = (f_o)_i \quad \text{auf} \quad C_i .$$

Umgekehrt erhält man durch die Vorgabe holomorpher T-Abbildungen $f_i : T \times C_i \to T \times \tilde{D}_{\tau(i)}$, welche die Bedingungen in (4.3.1) erfüllen, genau einen T-Morphismus $f : (X_T, A) \longrightarrow Y_T$ mit $f(0) = f_o$ und $\Psi_{\tau(i)T} f \Phi_{iT}^{-1} = f_{iT}$. Dies folgt unter Benutzung von [Ko]$_1$ (7.11) sowie (2.3), (2.4).

Wir definieren nun

$$E := \prod_{i \in I} \Gamma_h(C_i, (\varphi_i)_*(\mathcal{L}|Q_i))$$

und

$$F := \prod_{(i,j) \in I^2} \Gamma_h(C_{ij}, (\varphi_i)_*(\mathcal{L}|Q_i)) .$$

Dann sind E und F bzgl. der gegebenen Trivialisierungen von

$(\varphi_i)_*(\mathcal{L}|U_i(0))$ Hilbert-Räume und der induzierte Čech-Korandoperator $\delta : E \longrightarrow F$ ist stetig. Das Element $e_o := ((f_o)_i)_{i\in I}$ bzw. 0 betrachten wir fortan als ausgezeichneten Punkt von E bzw. F. Wir wollen uns als nächstes überlegen, daß durch die erste Bedingung in (4.3.1) im Falle $T = S$ ein analytischer S-Morphismus

$$\omega : S \times (E, e_o) \longrightarrow S \times (F, 0)$$

von banachanalytischen Raumkeimen definiert wird mit $T_{e_o}(\omega(0)) = \delta$. Dies ergibt sich mit Hilfe der folgenden etwas allgemeineren Betrachtung.

(4.4) Seien $U \subset\subset \mathbb{C}^\nu$ und $V \subset \mathbb{C}^\mu$ offene Teilmengen, $r : U \longrightarrow V$ eine holomorphe Abbildung mit $r(U) \subset\subset V$ und $S = (S,0)$ ein komplexer Raumkeim. Wir setzen

$$\mathrm{Hom}_S(S \times U, S \times V)^o := \{G \in \mathrm{Hom}_S(S \times U, S \times V) : G(0) = r\} .$$

Bezeichnet $\sigma_U : \Gamma_h(U)^\mu \times U \longrightarrow \mathbb{C}^\mu$ die Einsetzungsabbildung $(f,x) \longmapsto f(x)$, welche offenbar analytisch ist, so erhalten wir eine kanonische Abbildung

$$\text{(4.4.1)} \qquad \iota : \mathrm{Hom}(S, (\Gamma_h(U)^\mu, r)) \longrightarrow \mathrm{Hom}_S(S \times U, S \times V)^o$$
$$H \longmapsto \sigma_U(H \times \mathrm{id}_U) .$$

Ist $U' \subset\subset U$ relativ kompakt und offen, so gibt es umgekehrt eine kanonische Abbildung

$$\text{(4.4.2)} \qquad \varkappa : \mathrm{Hom}_S(S \times U,\ S \times V)^o \longrightarrow \mathrm{Hom}(S,\ (\Gamma_b(U')^\mu,\ r|U')) \quad {}^{*)}$$

mit $\varkappa\iota = \beta$, wobei $\beta : \mathrm{Hom}(S, (\Gamma_h(U)^\mu, r)) \longrightarrow \mathrm{Hom}(S, (\Gamma_b(U')^\mu, r))$ durch die Beschränkung $\Gamma_h(U) \longrightarrow \Gamma_b(U')$ induziert ist. Die Konstruktion von $\varkappa$ ergibt sich dabei sofort aus den beiden Identitäten

$$\mathrm{Hom}_S(S \times \overline{U},\ S \times \mathbb{C}) = \mathcal{O}_S \hat{\otimes}_{\mathbb{C}} \Gamma(\overline{U}') ,$$
$$\mathrm{Hom}(S,\ \Gamma_b(U')) = \mathcal{O}_S \hat{\otimes}_{\mathbb{C}} \Gamma_b(U') .$$

---

*) Hierbei bezeichnet $\Gamma_b(W)$ die Banachalgebra der beschränkten holomorphen Funktionen auf einer offenen Teilmenge $W \subset\subset \mathbb{C}^k$.

Man beachte, daß mit β auch ι injektiv ist.

Seien $p : T \longrightarrow S$ ein Morphismus komplexer Raumkeime und G bzw. H aus $\mathrm{Hom}_S(S\times U,\ S\times V)^o$ bzw. $\mathrm{Hom}(S,\ (\Gamma_h(U)^\mu,\ r))$. Dann gilt

$$\varkappa(G_T) = \varkappa(G)p$$

$$\iota(Hp) = \iota(H)_T \ .$$

Insbesondere sind ι und $\varkappa$ funktoriell in S .

Sei $s : V \longrightarrow W \subset \mathbb{C}^\rho$ eine weitere holomorphe Abbildung, wobei zusätzlich $V \subset\subset \mathbb{C}^\mu$ und $s(V) \subset\subset W$ gelte. Die übliche Komposition holomorpher Abbildungen induziert dann einen analytischen Morphismus zwischen Banachraumkeimen

$$K_{s,r} : (\Gamma_h(V)^\rho,\ s)\times(\Gamma_b(U)^\mu,\ r) \longrightarrow (\Gamma_h(U)^\rho,\ sr).$$

Die Taylorentwicklung holomorpher Funktionen auf V um den Aufpunkt r zusammen mit der Cauchy-Abschätzung der Ableitungen einer holomorphen Funktion liefert die Analytizität von $K_{s,r}$. In einem etwas veränderten Kontext wird diese Kompositionsabbildung ausführlich in [Bi-Ko] Kapitel II, §§ 6,8 behandelt. Erfüllt U die Kegelbedingung aus [Ad] p. 66, so kann man $\Gamma_b(U)$ durch $\Gamma_h(U)$ ersetzen. Ferner gilt für $G \in \mathrm{Hom}(S,(\Gamma_b(U)^\mu,r))$ und $H \in \mathrm{Hom}(S,(\Gamma_h(V)^\rho,s))$ die Formel

$$(4.4.4) \qquad \iota K_{s,r}\,(H\times G)\,\Delta_S = \iota(H)\ \iota(G)\ ,$$

wobei $\Delta_S : S \longrightarrow S\times S$ die Diagonalabbildung bezeichnet.

(4.5) Wir kehren wieder zur Situation in (4.3) zurück. Seien $\tilde{\Phi}_{ji}$ aus $\mathrm{Hom}(S,(\Gamma_b(C_{ij})^n,\ \varphi_{ji}))$ und $\tilde{\Psi}_{ji}$ aus $\mathrm{Hom}(S,(\Gamma_h(\tilde{D}_{\tau(i)})^m,\ \psi_{\tau(j)\tau(i)}))$, wobei wir abkürzend $\psi_{\alpha\beta}:=\Psi_{\alpha\beta}(0)$ gesetzt haben, mit $\iota(\tilde{\Phi}_{ji}) = \Phi_{ji}$ und $\iota(\tilde{\Psi}_{ji}) = \Psi_{\tau(j)\tau(i)}$. Wir definieren dann die S-Morhismen

$$M_{ij} : (\Gamma_h(C_j)^m, (f_o)_j) \times S \longrightarrow S \times \Gamma_h(C_{ij})^m ,$$

$$N_{ij} : S \times (\Gamma_h(C_i)^m, (f_o)_i) \longrightarrow S \times \Gamma_h(C_{ij})^m$$

durch die Komposition

$$pr_2\, M_{ij} := K_{(f_o)_j,\ \varphi_{ji}} \ (id \times \tilde{\Phi}_{ji})$$

bzw.

$$pr_2\, N_{ij} := K_{\psi_{\tau(j)\tau(i)}),(f_o)_i} (\tilde{\Psi}_{ji} \times id)\ (id_S \times \varepsilon_i)^{-1} ,$$

wobei $\varepsilon_i : \Gamma_h(D_i)^m \xrightarrow{\sim} \Gamma_h(C_i)^m$ der durch die Beschränkungsabbildung gegebene Isomorphismus ist (vgl. (2.4)). Für die Tangentialabbildungen über dem ausgezeichneten Punkt 0 von S gilt

$$T(M_{ij}(0))u = u(\varphi_{ij}) \qquad \text{für } u \in \Gamma_h(C_j)^m$$

bzw.

$$T(N_{ij}(0))v = D\psi_{ji}((f_o)_i)v \,|C_{ij} \qquad \text{für } v \in \Gamma_h(C_i)^m .$$

Wir setzen $M := (M_{ij})_{(i,j)\in I^2}$ und $N := (N_{ij})_{(i,j)\in I^2}$. Dann sind M und N S-Morphismen

$$S \times (E, e_o) \underset{N}{\overset{M}{\rightrightarrows}} S \times F .$$

Ferner sei $\omega : S \times (E, e_o) \longrightarrow S \times (F, 0)$ durch $pr_2\, \omega := pr_2(M-N)$ definiert und $Z := (pr_2 \omega)^{-1}(0)$. Es gilt dann $T\,\omega(0) = \hat{\delta}$. Der banachanalytische Raumkeim $Z = (Z, (0, e_o))$ hat nun folgende wichtige Eigenschaft:

<u>(4.6) Lemma.</u> Sei $p : T \longrightarrow S$ ein komplexer Raumkeim über S und $G : T \rightarrow S \times (E, e_o)$ ein S-Morphismus mit den Komponenten G_i für $i \in I$.

Dann erfüllt das System der $\iota(G_i) : T \times C_i \longrightarrow T \times \tilde{D}_{\tau(i)}$ die Bedingung (4.3.1) genau dann, wenn G über Z faktorisiert.

<u>Beweis.</u> Für die $\iota(G_i)$, $i \in I$, gilt (4.3.1) genau dann, wenn

$$\iota(G_j)\ \Phi_{jiT} = \Psi_{\tau(j)\tau(i)T}\ \iota(G_i)$$

erfüllt ist. Dies ist nach (4.4.4) gleichbedeutend mit

$$\iota K_{(f_o)_j, \varphi_{ji}} (G_j \times \tilde{\Phi}_{ji} p)\ \Delta_T = \iota K_{\psi_{\tau(j)\tau(i)}, (f_o)_i} (\tilde{\Psi}_{ji} p \times G_i) \Delta_T .$$

Da ι injektiv ist, ist die letzte Identität äquivalent mit

$$M_{ij}(G_j \times p)\ \Delta_T = N_{ij}\ (p \times G_i)\ \Delta_T \quad .$$

Dies zeigt die Behauptung.

<u>5. Beweis von (1.1) und (1.2)</u>

(5.1) Wir legen die Situation von Abschnitt (4.3) und (4.5) zugrunde und zeigen zunächst, daß der banachanalytische Raumkeim Z bereits endlichdimensional ist. Sei dazu $F' := \text{Bild}\ (T_{e_o}(\omega(0)))$. Dann ist F' nach (3.4) ein direkter Unterraum von F, es gibt also eine Projektion $p : F \longrightarrow F'$. Der Kern W des Doppelpfeils

$$S \times (E, e_o) \underset{(\mathrm{id} \times 0)}{\overset{(\mathrm{id} \times p)\omega}{\rightrightarrows}} S \times (F', 0)$$

ist nach [Dou], I Proposition 2(b), glatt über S mit $T_{e_o}(W(0)) = \text{Kern}\ (T_{e_o}(\omega(0)))$. Wegen $\dim_{\mathbb{C}} \text{Kern}\ (T_{e_o}(\omega(0))) < \infty$, vgl. (3.4), ist W endlichdimensional, damit aber auch Z, da $Z \subset W$ ein abgeschlossener Unterkeim ist.

Wir konstruieren nun einen Funktorisomorphismus

$$\eta_T : \mathrm{Hom}_T((X_T, A), Y_T)^o \longrightarrow \mathrm{Hom}_S(T, Z)$$

auf der Kategorie (Gan/S), der komplexen Raumkeime über S . Sei f aus $\mathrm{Hom}_T((X_T, A), Y_T)^{\circ}$ vorgegeben und $f_i : T \times C_i \longrightarrow T \times \tilde{D}_{\tau(i)}$ der induzierte T-Morphismus mit $f_i(0) = (f_o)_i$. Da nach Konstruktion f_i bereits auf einer Verdickung von C_i gegeben ist, gibt es ein eindeutig bestimmtes $\tilde{f}_i$ aus $\mathrm{Hom}(T,(\Gamma_h(C_i)^m, (f_o)_i))$ mit $\iota(\tilde{f}_i) = f_i$ für alle $i \in I$. Die Identitäten $f_j \Phi_{jiT} = \Psi_{\tau(j)\tau(i)T} f_i$ auf $T \times C_{ij}$, $(i,j) \in I^2$, sind dann nach (4.6) gleichwertig damit, daß $\tilde{f} := (\tilde{f}_i)_{i \in I}$ über Z faktorisiert. Wir setzen $\eta_T(f) := \tilde{f}$. Nach Konstruktion ist η ein Monomorphismus von Funktoren. Die Surjektivität ergibt sich aus (4.6) und (4.3), womit der Satz (1.1) bewiesen wäre.

(5.2) Nun zum Beweis von (1.2). Wir verwenden dazu das Darstellbarkeitskriterium von [S-V] Proposition (1.1). Die relative Darstellbarkeit (dies ist die erste Bedingung aus loc. cit.) folgt genauso, wie das in $[\mathrm{Ko}]_1$ Abschnitt 9 in einer etwas anderen Situation durchgeführt wurde, wobei sich hier noch einige Vereinfachungen ergeben. Wir verifizieren die Offenheit der Universalität für den nach (1.1) existierenden Abbildungskeim. Sei also $0 \in S$ ein Punkt, $f_o : (X(0), A(0)) \longrightarrow Y(0)$ eine feste holomorphe Abbildung und $f : (X_T, A_T) \longrightarrow Y_T$ die nach (1.1) gegebene universelle Deformation von f_o über dem Raumkeim $(T,0)$. Der Strukturmorphismus von $(T,0)$ sei mit $\rho : (T,0) \longrightarrow (S,0)$ bezeichnet. Wir fixieren ferner Meßatlanten $(\Phi_i)_{i \in I}$ bzw. $(\Psi_\alpha)_{\alpha \in \tilde{I}}$ für p bzw. q , wobei letzterer dem Polyederaxiom in jeder Faser genügt. Analog zu (4.1) führen wir die folgenden Bezeichnungen ein, wobei $i,j \in I$ und $s \in S$ sind

$$P_i := \Phi_i^{-1}(S \times D_i), \quad P_i^! := \Phi_i^{-1}(S \times D_i^!) ,$$

$$Q_i := P_i \cap B, \qquad Q_i^! := P_i^! \cap B' ,$$

$$C_{is} := \Phi_i(s)\,(Q_i(s)) ,$$

$$C_{ijs} := \Phi_i(s)\,(Q_i(s) \cap Q_j(s)) ,$$

$$C_{i\underline{j}s} := \Phi_i(s)\,(Q_i(s) \cap Q_j^!(s)) ,$$

$$D_{i\underline{j}} := \Phi_i(0)\,(P_i(0) \cap P_j^!(0)) .$$

Hierbei sind B und B' gemäß (2.5) definiert. Die Fasern $B(s)$ und $B'(s)$ sind dann offene Teilmengen von $X(s)$ mit glattem 1-konkaven Rand. Nach eventueller Verkleinerung von S um 0 liegt $\Phi_{ji}(s)\,(D_{ij})$ relativ kompakt in D_j für alle $s \in S$. Wir bezeichnen mit $\check{D}_{ijs}$ die Vereinigung derjenigen Zusammenhangskomponenten von D_{ij}, die mindestens einen Punkt von C_{ijs} enthalten. Durch kleines Verändern der Multiradien der D_i's kann man annehmen, daß $\check{D}_{ijs}$ unabhängig von $s \in S$ ist (wobei wiederum S zu schrumpfen ist) und wir schreiben dann $\check{D}_{ij}$ dafür. Es sei nun

$$\tilde{E} := \prod_{i\in I} \Gamma_h(D_i, \mathcal{O}^m_{\mathbb{C}^n}) ,$$

$$E_s := \prod_{i\in I} \Gamma_h(C_{is}, \mathcal{O}^m_{\mathbb{C}^n}) , \qquad \text{für } s \in S,$$

$$\tilde{F} := \prod_{(i,j)\in I^2} \Gamma_h (\check{D}_{ij}, \mathcal{O}^m_{\mathbb{C}^n}),$$

$$F_s := \prod_{(i,j)\in I^2} \Gamma_h(C_{ijs}, \mathcal{O}^m_{\mathbb{C}^n}), \qquad \text{für } s \in S .$$

Man hat dann die Beschränkungsabbildungen

$$\beta_s : \tilde{E} \longrightarrow E_s ,$$

$$\gamma_s : \tilde{F} \longrightarrow F_s ,$$

für jedes $s \in S$, wobei β_s ein Isomorphismus ist (vgl. (2.4)) und jedes γ_s injektiv nach Definition von $\check{D}_{ij}$.

Ohne Einschränkung dürfen wir ferner annehmen, daß es eine Abbildung $\tau : I \longrightarrow \tilde{I}$ gibt mit $f(t)\,(P_i(\rho(t))) \subset\subset \tilde{P}_{\tau(i)}\,(\rho(t))$ für alle $t \in T$, wobei $\tilde{P}_\alpha = \Psi_\alpha^{-1}(S \times \tilde{D}_\alpha)$ mit $\alpha \in \tilde{I}$ ist. Somit definiert jedes $f(t)$ in kanonischer Weise ein Element e_t aus $\tilde{E}$. Die Komposition mit $\Phi_{ji}(s)$ bzw. $\Psi_{\tau(j)\tau(i)}(s)$, $s \in S$, liefert dann gemäß der Konstruktion von (4.5) eine holomorphe Abbildung

$$\tilde{\omega} : S \times U \longrightarrow S \times \tilde{F}$$

von banachanalytischen Räumen über S, wobei U eine offene Umgebung von e_o in E ist. Die im Beweis von (1.1) angegebene Konstruktion von $(T,0)$ zeigt

$$(T,0) = ([pr_2(id_S \times \gamma_o)\tilde{\omega}]^{-1}(0),\ (0,e_o))$$

Wir können daher annehmen, daß $T \subset S \times U$ ein abgeschlossener Unterraum ist. Aus (5.3) und (3.4) folgt, daß die Tangentialabbildung $T_{e_o}\tilde{\omega}(0) : \tilde{E} \longrightarrow \tilde{F}$ direkt mit endlich dimensionalem Kern ist. Setzen wir $(Z,0) := ([pr_2\tilde{\omega}]^{-1}(0),\ (0,e_o))$, so ist $(Z,0)$ ein komplexer Raumkeim mit $(Z,0) \subset (T,0)$. Wegen (5.4) muß aber notwendigerweise $(Z,0) = (T,0)$ sein und nach Verkleinern der Situation also $Z = T$.

Sei jetzt $t \in T$ ein beliebiger Punkt. Wir betrachten das folgende kommutative Diagramm

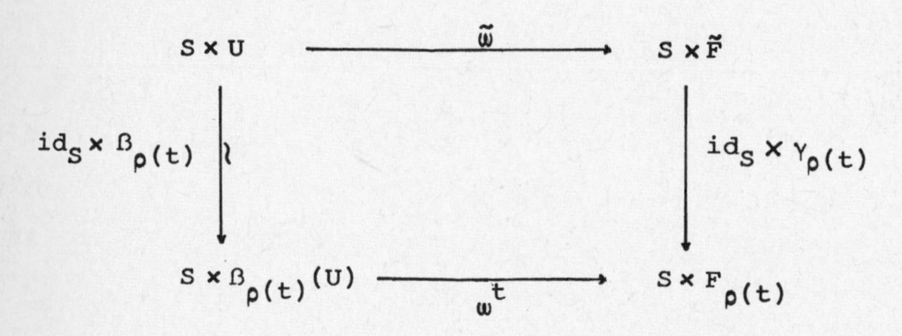

von banachanalytischen Räumen über S. Es ist $\tilde{\omega}(\rho(t),e_t)=(\rho(t),0)$ für jedes $t \in T$ und $id_S \times \beta_{\rho(t)}$ induziert einen Morphismus

$$T \longrightarrow [pr_2\omega^t]^{-1}(0) .$$

Die Behauptung ist nach Konstruktion der universellen Deformation von $f(t)$ gezeigt, falls dieser Morphismus als Keim im Punkt $(\rho(t),e_t)$ ein Isomorphismus ist. Dies folgt aber sofort, ähnlich wie eben, aus (5.4) und (3.4) angewandt auf $T_{e_t}\omega^t(\rho(t))$.

Die beiden folgenden Aussagen wurden im Beweis von (1.2) benutzt.

<u>(5.3) Lemma.</u> Seien $\alpha : E \longrightarrow F$ und $\beta : F \longrightarrow G$ stetige $\mathbb{C}$-lineare Abbildungen von Banachräumen. Ist β injektiv, so ist mit $\beta\alpha$ auch α direkt.

Der Beweis ist trivial.

(5.4) Lemma. Sei W ein komplexer Raumkeim und $i : F \longrightarrow F_o$ eine injektive, stetige $\mathbb{C}$-lineare Abbildung von Banachräumen. Dann ist auch die kanonische Abbildung

$$\begin{aligned} \mathrm{Hom}(W,F) &\longrightarrow \mathrm{Hom}(W, F_o) , \\ \varphi &\longmapsto i\varphi \end{aligned}$$

injektiv.

Der Beweis folgt unmittelbar aus der Aussage (II.12.4) von [Bi-Ko].

6. Eine Vergleichsaussage.

Sei A eine kompakte komplexe Mannigfaltigkeit und $\pi : E \longrightarrow A$ ein holomorphes Vektorraumbündel auf A. Wir nennen E 1-positiv, falls der Nullschnitt von E eine relativ kompakte offene Umgebung B mit glattem 1-konkavem Rand ∂B besitzt. Diese Definition der 1-Positivität ist leicht allgemeiner wie die in [Co-Gr] angegebene, doch reicht dies für unseren Zweck aus.

(6.1) Satz. Sei $\pi : E \longrightarrow A$ ein 1-positives Vektorraumbündel auf der kompakten komplexen Mannigfaltigkeit A und $\mathcal{L}$ ein kohärenter lokalfreier $\mathcal{O}_A$ - Modul sowie

$$r^k(\pi^*\mathcal{L}) : \varinjlim_{W \supset A} H^k(W, \pi^*\mathcal{L}) \longrightarrow H^k(\hat{E}, \hat{\pi}^*\mathcal{L})$$

der kanonische Homomorphismus, wobei der Limes über alle offenen Umgebungen W von A in E genommen wird.

Dann ist $r^k(\pi^*\mathcal{L})$ bijektiv für $k = 0$ und injektiv für $k = 1$.

(6.2) Korollar. Ersetzt man in der Situation von (6.1) die Garbe $\pi^*\mathcal{L}$ durch die Garbe der holomorphen Vektorfelder Θ_E auf E, so ist $r^k(\Theta_E)$ bijektiv für $k = 0$ und injektiv für $k = 1$.

Der Beweis von (6.2) folgt sofort aus (6.1) mit Hilfe der kurzen exakten Sequenz

$$0 \longrightarrow \pi^*\mathcal{E} \longrightarrow \Theta_E \longrightarrow \pi^*\Theta_A \longrightarrow 0\,,$$

vgl. $[Ko]_2$ (4.3), wobei $\mathcal{E}$ die Garbe der holomorphen Schnitte von $E \longrightarrow A$ bezeichnet.

Zum Beweis von (6.1) benutzen wir die folgende Aussage.

(6.3) Proposition. Sei $\pi : E \longrightarrow A$ ein 1-positives Vektorraumbündel und $\mathcal{L}$ ein kohärenter lokalfreier $\mathcal{O}_A$ - Modul.

Dann gibt es zu jeder offenen Umgebung W des Nullschnitts A von E eine offene Teilmenge $W' \subset W$ mit $A \subset W'$, so daß die Inklusion

$$\text{Kern}\,(H^1(W,\pi^*\mathcal{L}) \longrightarrow H^1(\hat{E},\hat{\pi}^*\mathcal{L})) \subset$$
$$\text{Kern}\,(H^1(W,\pi^*\mathcal{L}) \xrightarrow[\text{Beschr.}]{} H^1(W',\pi^*\mathcal{L}))$$

gilt.

Beweis. Wir fixieren zunächst eine endliche offene Steinsche Überdeckung $(A_i)_{i\in I}$ von A zusammen mit Karten $\varphi_i : A_i \xrightarrow{\sim} V_i \subset\subset \mathbb{C}^n$, $n = \dim A$, und Trivialisierungen $\Phi_i : E_{A_i} \xrightarrow{\sim} A_i \times \mathbb{C}^r$, $r = \mathrm{rg}(E)$, von E und ebenso von $\mathcal{L}$, so daß die Übergangsmatrizen von E und $\mathcal{L}$ beschränkt sind. Ferner sei $D \subset \mathbb{C}^r$ ein offener beschränkter Polyzylinder um 0 mit konzentrischen Schrumpfungen $D' \subset\subset D^* \subset\subset D$. Wir setzen $U_i := \Phi_i^{-1}(A_i \times D)$, $U_i^* := \Phi_i^{-1}(A_i \times D^*)$, $U_i' := \Phi_i^{-1}(A_i \times D')$ und $\mathfrak{U} := (U_i)_{i\in I}$, $\mathfrak{U}^* := (U_i^*)_{i\in I}$ sowie $\mathfrak{U}' := (U_i')_{i\in I}$. Ist D^* genügend klein, so gibt es eine offene Teilmenge B von E mit 1-konkavem Rand ∂B und mit $|\mathfrak{U}^*| \subset\subset B \subset\subset |\mathfrak{U}| \subset\subset W$. Es sei $W' := |\mathfrak{U}'|$. Wir füllen $\mathfrak{U}^*$ zu einer endlichen offenen Überdeckung $\mathfrak{W}$ von B auf, die eine Verfeinerung von $\mathfrak{U}$ ist. Dann haben wir bzgl. der gegebenen Karten die Beschränkungsabbildungen

$$Z^1_{L^2}(\mathfrak{U},\pi^*\mathcal{L}) \to Z^1_{L^2}(\mathfrak{W},\pi^*\mathcal{L}) \to Z^1_{L^2}(\mathfrak{U}^*,\pi^*\mathcal{L}) \to Z^1_{L^2}(\mathfrak{U}',\pi^*\mathcal{L})$$

von Hilbert-Räumen. Jedes Element z aus $Z^1(\mathfrak{U},\pi^*\mathcal{L})$ besitzt eine kanonische homogene Entwicklung $z = \sum_{e\in\mathbb{N}} z_e$ mit

$z_e \in Z^1(\mathfrak{U} \cap A, S^e(\mathcal{E}^\vee) \otimes_{\mathcal{O}_A} \mathcal{L})$ längs den Fasern von π. Sei überdies z aus $Z^1_{L^2}(\mathfrak{U}, \pi^*\mathcal{L})$. Wir dürfen annehmen, daß diese Entwicklung bereits auf $\mathfrak{U}$ konvergiert und daß es eine Konstante $C_1 > 0$ gibt mit $\|z_e\|_{\mathfrak{U}} \leq C_1 \|z\|_{\mathfrak{U}}$ für alle $e \in \mathbb{N}$. Zerfällt nun z formal, so gibt es zu jedem $e \in \mathbb{N}$ ein $a_e \in C^0(\mathfrak{U} \cap A, S^e(\mathcal{E}^\vee) \otimes_{\mathcal{O}_A} \mathcal{L})$ mit $z_e = \delta a_e$. Benutzt man jetzt die Lösung des $\bar{\partial}$ - Problems mit Abschätzung nach $[\text{Hö}]_1$ wie dies bereits im Beweis von (3.3) vorgeführt wurde, so gibt es eine Konstante $C_2 > 0$ und zu jedem $e \in \mathbb{N}$ ein $c(e) \in C^0_{L^2}(\mathfrak{W}, \pi^*\mathcal{L})$ mit $z_e|\mathfrak{W} = \delta\, c(e)$ und $\|c(e)\|_{\mathfrak{W}} \leq C_2 \|z_e\|$. Sei nun $c_e := [c(e)|\mathfrak{U}^*]_e$ der homogene Bestandteil vom Grad e von $c(e)|\mathfrak{U}^*$ und $c := \sum_{e \in \mathbb{N}} c_e$. Dann folgt $z|\mathfrak{U}^* = \delta c$ und $\|c_e\|_{\mathfrak{U}^*} \leq C_3 \|c(e)\|_{\mathfrak{W}}$ für alle $e \in \mathbb{N}$ mit einer Konstanten $C_3 > 0$. Nach dem Schwarz'schen Lemma folgt $\|c_e\|_{\mathfrak{U}'} \leq C_4 \|c_e\|_{\mathfrak{U}^*} q^e$ für ein geeignetes $q \in \,]0,1[$ und für alle $e \in \mathbb{N}$. Somit liegt c in $C^0_{L^2}(\mathfrak{U}', \pi^*\mathcal{L})$ und es gilt $z|\mathfrak{U}' = \delta c$. Dies zeigt die Behauptung.

<u>Beweis</u> von (6.1). Die Injektivität von $r^1(\pi^*\mathcal{L})$ ergibt sich sofort aus (6.3). Zum Beweis der Bijektivität von $r^0(\pi^*\mathcal{L})$ betrachten wir das folgende kanonische kommutative Diagramm für eine offene Umgebung W von A in E

$$\begin{array}{ccc}
\coprod_{e \in \mathbb{N}} H^k(A, S^e(\mathcal{E}^\vee) \otimes_{\mathcal{O}_A} \mathcal{L}) & \xrightarrow{\alpha_W} & H^k(W, \pi^*\mathcal{L}) \\
\downarrow i & & \downarrow r^k(\pi^*\mathcal{L}; W) \\
\prod_{e \in \mathbb{N}} H^k(A, S^e(\mathcal{E}^\vee) \otimes_{\mathcal{O}_A} \mathcal{L}) & \xrightarrow[\hat{\alpha}]{} & H^k(\hat{E}, \widehat{\pi^*\mathcal{L}})
\end{array}$$

Offenbar ist $\hat{\alpha}$ ein Isomorphismus und daher α_W injektiv. Sei nun $k = 0$. Dann ist für $W = B$ mit 1-konkavem Rand ∂B die Kohomologiegruppe $H^0(B, \pi^*\mathcal{L})$ nach dem Beweis von (3.3) endlichdimensional. Somit gilt $H^0(A, S^e(\mathcal{E}^\vee) \otimes_{\mathcal{O}_A} \mathcal{L}) = 0$ für alle $e \gg 0$ und folglich ist i bijektiv. Dies liefert die Surjektivität von $r^o(\pi^*\mathcal{L}; B)$. Die Injektivität von $r^o(\pi^*\mathcal{L})$ ist klar.

Literatur

[Ad] Adams, R. Sobolev Spaces. Academic Press, London 1975

[An-Gr] Andreotti, A., Grauert, H. Théorèmes de finitude pour la cohomologie des espaces complexes. Bull. Soc. math. France 90, 193 - 259 (1962)

[Bi-Ko] Bingener, J., Kosarew, S. Lokale Modulräume in der komplexen Analysis. Manuscript, Regensburg 1986

[Co-Gr] Commichau, M., Grauert, H. Das formale Prinzip für kompakte komplexe Untermannigfaltigkeiten mit 1-positivem Normalenbündel. In "Recent Developments in Several Complex Variables", Ann. Math. Stud. 100, 101 - 126 (1981)

[Dou] Douady, A. Le problème des modules locaux pour les espaces $\mathbb{C}$-analytiques compacts. Ann. scient. Éc. Norm. Sup. 4e sér. 7, 569 - 602 (1974)

[Gu-Ro] Gunning, R., Rossi, H. Analytic functions of several complex variables. Prentice-Hall, Englewood Cliffs, N.J., 1965

[Hö]$_1$ Hörmander, L. L^2 - Estimates and existence theorems for the $\bar{\partial}$ operator. Acta. Math. 113, 89-152 (1965)

[Hö]$_2$ Hörmander, L. An Introduction to Complex Analysis in Several Variables. North Holland P.C., Amsterdam 1973

[Ko]$_1$ Kosarew, S. Modulräume holomorpher Abbildungen auf konkaven komplexen Räumen. In Vorbereitung: Ann. Sc. École Norm. Sup.

[Ko]$_2$ Kosarew, S. Ein allgemeines Kriterium für das formale Prinzip.
Preprint, Göttingen 1986

[S-V] Schuster, H.W., Vogt, A. The moduli of quotients of a compact complex space.
Preprint 1985.

Siegmund Kosarew
Mathematisches Institut der
Universität, Sonderforschungs-
bereich 170 "Geometrie und Analysis"
Bunsenstr. 3-5
D-3400 Göttingen
Bundesrepublik Deutschland

Stable rationality of some moduli spaces of vector bundles on P^2

Masaki Maruyama

Introduction.

Let k be an algebraically closed field. It is known that all the moduli spaces of stable vector bundles on $P = P_k^2$ are irreducible, non-singular and quasi-projective ([4], [2], [1]). Restricting ourselves to the case where the first Chern class is zero, we set $M'(r, n)$ (or, $M'(r, n)^{\mu}$) to be the moduli space of stable (or, μ-stable, resp.) vector bundles E with the following properties:

(a) $c_1(E) = 0$, $c_2(E) = n$, and the rank $r(E)$ of E is r,
(b) for general line ℓ in P, $E|_{\ell} \simeq \mathcal{O}_{\ell}^{\oplus r}$.

Then, for $2 \leq r \leq n$, both $M'(r, n)$ and $M'(r, n)^{\mu}$ are non-empty open subschemes of the moduli space $M(r, n)$ of stable vector bundles with the invariants in (a) and moreover, the first contains the second. In [5] and [6] the following are proved.

Theorem 0.1. (1) *There are subschemes* $Z(r, n)$ $(2 \leq r \leq n)$ *in* $M'(n, n)$ *such that* (i) $M'(r, n)^{\mu}$ *is isomorphic to* $Z(r, n)$, (ii) $M'(n, n) = \coprod_{2\leq r\leq n} Z(r, n)$ *and* (iii) $\overline{Z(r, n)} = \coprod_{s\leq r} Z(s, n)$, *where* $\overline{}$ *means the closure in* $M'(n, n)$.

(2) *There is an over field* K *of the function field* L *of* $M(r, n)$ *such that* (i) K *is purely transcendental over* k, (ii) K *is a Galois extension of* L *with Galois group* S_n.

In this article we shall prove that for all $n \geq 2$, the moduli space $M(n, n)$ is stably rational, that is, for a positive integer N, $M(n, n) \times_k P_k^N$ is a rational variety. This and the theorem stated in the above seem to give a strong evidence of the following conjecture.

Conjecture 0.2. All the moduli spaces of stable vector bundles on P are stably rational.

Throughout this paper a vector bundle means a locally free sheaf of finite, constant rank and P denotes the projective plane over an algebraically closed field k. The set of k-rational points of a k-scheme X is denoted by $X(k)$. If $f : X \longrightarrow Y$ be a morphism of schemes and F is a coherent sheaf on X, then $X(y)$ denotes the fibre of f over a point y of Y and $F(y)$ does $F \otimes_{\mathcal{O}_X} \mathcal{O}_{X(y)}$. Especially, for a point x of X, $F(x)$ means the fibre $F \otimes_{\mathcal{O}_X} k(x)$ of F over x.

§1. A universal family of some extensions.

Let us start with a lemma which shows the structure of vector bundles in $M'(n, n)$.

Lemma 1.1. *Let E be a vector bundle on P with the following properties*

(a) $c_1(E) = 0$ *and* $c_2(E) = r(E) = n$,

(b) $H^0(P, E) = 0$.

If x is a k-rational point of P, then $V(E, x) = \{s \in H^0(P, E(1)) \mid s(x) = 0\}$ is a linear subspace of dimension n in $H^0(P, E(1))$. Assume further that there are exactly n lines $\ell_1, \ldots, \ell_n$ such that every ℓ_i contains x and $E|_{\ell_i} \not\simeq \mathcal{O}_{\ell_i}^{\oplus n}$. Then the cokernel $Q(E, x)$ of the natural map $V(E, x) \otimes_k \mathcal{O}_P \longrightarrow E(1)$ is isomorphic to $\bigoplus_{i=1}^{n} \mathcal{O}_{\ell_i}$.

Proof. Since E is 1-regular ([5] Lemma 1.8), both $H^1(P, E(1))$ and $H^2(P, E(1))$ vanish and then $h^0(P, E(1)) = 2n$ by Riemann-Roch Theorem. $V(E, x)$ is the kernel of the restriction map $H^0(P, E(1)) \longrightarrow H^0(P, E(1)(x)) \simeq k^{\oplus n}$. Since $E(1)$ is generated by $H^0(P, E(1))$, so is $E(1)(x)$ and hence the above map is surjective. Thus $\dim V(E, x) = n$. To prove the latter half, take a line ℓ passing through x. Then $E(1)|_\ell \simeq \mathcal{O}^{\oplus r} \oplus (\bigoplus_{j=1}^{n-r} \mathcal{O}_\ell(\alpha_j))$ with $\alpha_j > 0$ and for general ℓ, $r = 0$. Since $H^0(P, E) = H^1(P, E) = 0$, the natural map $H^0(P, E(1)) \longrightarrow H^0(P, E(1)|_\ell)$ is bijective. Thus the image of $V(E, x)$ to $H^0(\ell, E(1)|_\ell)$ is $\bigoplus_{i=1}^{n-r} \{s \in H^0(\ell, \mathcal{O}_\ell(\alpha_i)) \mid s(x) = 0\}$. We see therefore that $Q(E, x)|_\ell \simeq \mathcal{O}_\ell^{\oplus r} \oplus k(x)^{\oplus(n-r)}$. If $E(1)|_{\ell_i} \simeq \mathcal{O}_{\ell_i}^{\oplus r_i} \oplus (\bigoplus_{\alpha > 0} \mathcal{O}_{\ell_i}(\alpha))$, then we have a surjective $\rho_i : Q(E, x) \longrightarrow \mathcal{O}_{\ell_i}^{\oplus r_i}$. We shall prove that $\rho = \bigoplus_{i=1}^{m} \rho_i : Q(E, x) \longrightarrow \bigoplus_{i=1}^{m} \mathcal{O}_{\ell_i}^{\oplus r_i}$ is an isomorphism, which completes the proof. The above argument first shows that ρ is surjective on $P - x$ and $\mathrm{Supp}(Q(E, x)) = \bigcup_{i=1}^{n} \ell_i$. Thus there is an exact sequence

$$0 \longrightarrow V(E, x) \otimes_k \mathcal{O}_P \longrightarrow E(1) \longrightarrow Q(E, x) \longrightarrow 0$$

and hence $c_1(Q(E, x)) = n$. These imply that ρ is isomorphic outside a finite set of points and $r_i = 1$ for all i. From the above exact sequence we deduce that no local sections of $Q(E, x)$ are supported by a finite set of points. Hence ρ is injective. On the other hand, the above exact sequence again shows that $c_2(Q(E, x)) = c_2(E(1)) = n(n+1)/2$ which is equal to $c_2(\bigoplus_{i=1}^{n} \mathcal{O}_{\ell_i})$. Therefore, ρ is an isomorphism. Q. E. D.

The assumption for the latter half in the above lemma means that in the dual plane the curve of jumping lines of E intersects transversely the line formed by lines in P passing through x. This holds for the vector bundles E which correspond to general points of $M(n, n)$. And then we have an exact sequence

$$0 \longrightarrow \mathcal{O}_P^{\oplus n} \xrightarrow{\lambda} E(1) \xrightarrow{\mu} \bigoplus_{i=1}^{n} \mathcal{O}_{\ell_i} = Q \longrightarrow 0,$$

where ℓ_i's are mutually distinct lines passing through the point x. Conversely, assume that E fits in the above exact sequence. Restricting the sequence to ℓ_i, we have a surjection

$$E(1)|_{\ell_i} \longrightarrow \mathcal{O}_{\ell_i} \oplus k(x)^{\oplus(n-1)} \longrightarrow \mathcal{O}_{\ell_i}$$

Then we see easily that ℓ_i is a jumping line of E passing through x. Since $Q(x) = k(x)^{\oplus n}$ and $k(x)^{\oplus n} \simeq E(x) \longrightarrow Q(x)$ is surjective, the subspace $H^0(P, \mathcal{O}_P^{\oplus n})$ in $H^0(P, E(1))$ is contained in $V(E, x)$ and Lemma 1.1 implies that they coincide. The argument in the proof of Lemma 1.1 shows that if a line ℓ passing through x is a jumping line, then ℓ is contained in $\mathrm{Supp}(Q)$. Thus $\{\ell_1, \ldots, \ell_n\}$ is the set of jumping lines passing through x. Moreover, we have the uniqueness of the above exact sequence, that is, if we have another exact sequence

$$0 \longrightarrow \mathcal{O}_P^{\oplus n} \xrightarrow{\lambda'} E(1) \xrightarrow{\mu'} \bigoplus_{i=1}^{n} \mathcal{O}_{\ell'_i} \longrightarrow 0.$$

with ℓ'_i mutually distinct lines passing through x. Then $\{\ell_1, \ldots, \ell_n\} = \{\ell'_1, \ldots, \ell'_n\}$ and there is a commutative diagram:

$$\begin{array}{ccccccccc}
0 & \longrightarrow & \mathcal{O}_P^{\oplus n} & \xrightarrow{\lambda} & E(1) & \xrightarrow{\mu} & \bigoplus_{i=1}^{n} \mathcal{O}_{\ell_i} = Q & \longrightarrow & 0 \\
 & & \downarrow\wr & & \| & & \downarrow\wr & & \\
0 & \longrightarrow & \mathcal{O}_P^{\oplus n} & \xrightarrow{\lambda'} & E(1) & \xrightarrow{\mu'} & \bigoplus_{i=1}^{n} \mathcal{O}_{\ell'_i} & \longrightarrow & 0
\end{array}$$

Thus we obtain the following

(1.2) <u>For a k-rational point</u> x <u>of</u> P, <u>there is a non-empty open set</u> $U(x)$ <u>of</u> $M(n, n)$ <u>such that</u> y <u>of</u> $M(n, n)(k)$ <u>is in</u> $U(x)(k)$ <u>if and only if</u> y <u>corresponds to a vector bundle</u> E <u>whose jumping lines passing through</u> x <u>consist of mutually distinct</u> n <u>lines or equivalently</u>, E <u>fits in an exact sequence of the following form</u>

$$0 \longrightarrow \mathcal{O}_P^{\oplus n} \longrightarrow E(1) \longrightarrow \bigoplus_{i=1}^{n} \mathcal{O}_{\ell_i} \longrightarrow 0,$$

<u>where</u> ℓ_i's <u>are mutually distinct lines passing through the point</u> x. <u>Moreover, the above exact sequence is unique up to an isomorphism.</u>

The set of lines in P passing through x is parametrized by a projective line $L \simeq P_k^1$ and furthermore we have a universal family $\tilde{\ell} \subset L\times P$ of such lines. Let $\tilde{Y}_n$ be the open set $\{(y_1, \ldots, y_n) \in L\times\cdots\times L \mid y_i \neq y_j \text{ if } i \neq j\}$ in $L\times\cdots\times L$. The symmetric group S_n acts on $\tilde{Y}_n$ freely and the quotient Y_n is an open subscheme of the symmetric product of n copies of L. For the i-th projection p_i of $\tilde{Y}_n$ to L, $p_i{}^*(\tilde{\ell})$ is a relative Cartier divisor on $\tilde{Y}_n\times P$ over $\tilde{Y}_n$. S_n acts on $\tilde{\ell}(n) = \sum_{i=1}^{n} p_i{}^*(\tilde{\ell})$ and $\tilde{Q}_n = \bigoplus_{i=1}^{n} p_i{}^*(\mathcal{O}_{\tilde{\ell}})$ compatibly with the action on $\tilde{Y}_n$. Thus, by the descent theory, we have a relative Cartier divisor $\ell(n)$ on $Y_n\times P$ over Y_n and a Y_n-flat coherent sheaf Q_n on $Y_n\times P$ whose pull-back to $\tilde{Y}_n\times P$ are $\tilde{\ell}(n)$ and $\tilde{Q}_n$, respectively. $\ell(n)$ parametrizes bijectively all the unions of mutually distinct n lines in P passing through x and for every $y = (y_1, \ldots, y_n)$ in $Y_n(k)$, $Q_n(y) \simeq \bigoplus_{i=1}^{n} \mathcal{O}_{\tilde{\ell}(y_i)}$. Moreover, by using the descent theory, we can easily see the following universal property for the $(Y_n, \ell(n), Q_n)$.

(1.3) <u>Let</u> T <u>be an effective relative Cartier divisor on</u> $S\times P$ <u>over</u> S <u>and</u> R <u>be an</u> S-<u>flat coherent</u> $\mathcal{O}_T$-<u>module</u>. <u>Assume that for every</u> z <u>in</u> $S(k)$, $T(z)$ <u>is the union of mutually distinct</u> n <u>lines</u> $\ell_1, \ldots, \ell_n$ <u>in</u> P <u>passing through</u> x <u>and</u> $R(z) \simeq \bigoplus_{i=1}^{n} \mathcal{O}_{\ell_i}$. <u>Then, for every</u> z <u>in</u> $S(k)$, <u>there are an open neighborhood</u> U <u>of</u> z <u>and a morphism</u> $f : U \longrightarrow Y_n$ <u>such that</u> $T_U \simeq (f\times 1)^*(\ell(n))$ <u>and</u> $R_U \simeq (f\times 1)^*(Q_n)$.

<u>Lemma</u> 1.4. <u>For simplicity we set</u> $P' = Y_n\times P$.

(1) $\mathcal{E}xt^1_{\mathcal{O}_{P'}}(Q_n, \mathcal{O}_{P'}^{\oplus n})$ <u>is flat over</u> Y_n.

(2) <u>For every morphism</u> $f : S \longrightarrow Y_n$ <u>of algebraic</u> k-<u>schemes, the natural homomorphism</u> $(f\times 1)^*(\mathcal{E}xt^1_{\mathcal{O}_{P'}}(Q_n, \mathcal{O}_{P'}^{\oplus n})) \longrightarrow \mathcal{E}xt^1_{\mathcal{O}_{S\times P}}((f\times 1)^*(Q_n),$

$\mathcal{O}_{S\times P}^{\oplus n}$) <u>is an isomorphism</u>.

(3) <u>For the projection</u> $p : Y_n\times P \longrightarrow Y_n$, $p_*(\mathcal{E}xt^1_{\mathcal{O}_{P'}}(Q_n, \mathcal{O}_{P'}^{\oplus n}))$ <u>is a locally free sheaf on</u> Y_n <u>of rank</u> $2n^2$. <u>Moreover, for the</u> f <u>in</u> (2), <u>the natural homomorphism</u> $f^*p_*(\mathcal{E}xt^1_{\mathcal{O}_{P'}}(Q_n, \mathcal{O}_{P'}^{\oplus n})) \longrightarrow p'_*(\mathcal{E}xt^1_{\mathcal{O}_{S\times P}}((f\times 1)^*(Q_n), \mathcal{O}_{S\times P}^{\oplus n}))$ <u>is isomorphic, where</u> $p' : S\times P \longrightarrow S$ <u>is the first projection</u>.

(4) <u>For the</u> f <u>in</u> (2), $\mathcal{H}om_{\mathcal{O}_{S\times P}}((f\times 1)^*(Q_n), \mathcal{O}_{S\times P}^{\oplus n}) = 0$.

<u>Proof</u>. Since Q_n is flat over Y_n, $\operatorname{depth}_y Q_n = n+1$ at every y in $P'(k)$. Thus, there is a resolution of length 1 by locally free coherent sheaves:

$$0 \longrightarrow E_1 \longrightarrow E_0 \longrightarrow Q_n \longrightarrow 0.$$

The flatness of Q_n implies that for the f in (2), the following sequence is exact,too

$$0 \longrightarrow (f\times 1)^*(E_1) \longrightarrow (f\times 1)^*(E_0) \longrightarrow (f\times 1)^*(Q_n) \longrightarrow 0.$$

We have therefore the following exact commutative diagram

$$\begin{array}{ccccc} & & (f\times 1)^*(\mathcal{H}om_{\mathcal{O}_{P'}}(E_0, \mathcal{O}_{P'}^{\oplus n})) & \xrightarrow{u} & \\ & & \downarrow & & \\ 0 \longrightarrow \mathcal{H}om_{\mathcal{O}_{P_S}}((f\times 1)^*(Q_n), \mathcal{O}_{P_S}^{\oplus n}) & \longrightarrow & \mathcal{H}om_{\mathcal{O}_{P_S}}((f\times 1)^*(E_0), \mathcal{O}_{P_S}^{\oplus n}) & \xrightarrow{v} & \end{array}$$

$$\begin{array}{ccccc} (f\times 1)^*(\mathcal{H}om_{\mathcal{O}_{P'}}(E_1, \mathcal{O}_{P'}^{\oplus n})) & \longrightarrow & (f\times 1)^*(\mathcal{E}xt^1_{\mathcal{O}_{P'}}(Q_n, \mathcal{O}_{P'}^{\oplus n})) & \longrightarrow & 0 \\ \downarrow & & \downarrow & & \\ \mathcal{H}om_{\mathcal{O}_{P_S}}((f\times 1)^*(E_1), \mathcal{O}_{P_S}^{\oplus n}) & \longrightarrow & \mathcal{E}xt^1_{\mathcal{O}_{P_S}}((f\times 1)^*(Q_n), \mathcal{O}_{P_S}^{\oplus n}) & \longrightarrow & 0 \end{array}$$

Since the first and the second vertical arrows are isomorphic, so is the third, which proves (2). When $S = \operatorname{Spec}(k)$, $\mathcal{H}om_{\mathcal{O}_{P_S}}((f\times 1)^*(Q_n), \mathcal{O}_{P_S}^{\oplus n}) = 0$ and hence v is injective. This and the above exact sequence imply that $u(s)$ is injective for all s in $S(k)$. We deduce from this that $(f\times 1)^*(\mathcal{E}xt^1_{\mathcal{O}_{P'}}(Q_n, \mathcal{O}_{P'}^{\oplus n}))$ is flat over S and u is injective for every f (see, for example, [4] Lemma 6.5). Therefore, v is injective or equivalently, $\mathcal{H}om_{\mathcal{O}_{P_S}}((f\times 1)^*(Q_n), \mathcal{O}_{P_S}^{\oplus n}) = 0$. It remains to prove (3). Pick a point y of $Y_n(k)$. Then $Q(y) \simeq \bigoplus_{i=1}^{n} \mathcal{O}_{\ell_i}$ with

$\ell_1, \ldots, \ell_n$ mutually distinct lines passing through x. Since $\mathcal{E}xt^1_{\mathcal{O}_P}(\mathcal{O}_{\ell_i}, \mathcal{O}_P) \simeq \omega_{\ell_i}(3) \simeq \mathcal{O}_{\ell_i}(1)$, we see that $\mathcal{E}xt^1_{\mathcal{O}_P}(Q_n(y), \mathcal{O}_P^{\oplus n}) \simeq \bigoplus_{i=1}^{n} \mathcal{O}_{\ell_i}(1)^{\oplus n}$. This and (2) show that $h^0(P, \mathcal{E}xt^1_{\mathcal{O}_{P'}}(Q_n, \mathcal{O}_{P'}^{\oplus n})(y)) = 2n^2$ and $H^j(P, \mathcal{E}xt^1_{\mathcal{O}_{P'}}(Q_n, \mathcal{O}_{P'}^{\oplus n})(y)) = 0$ if $j > 0$. Since $\mathcal{E}xt^1_{\mathcal{O}_{P'}}(Q_n, \mathcal{O}_{P'}^{\oplus n})$ is flat over Y_n, the base change theorem yields (3). Q. E. D.

By virtue of (3) of the above lemma, $F = p_*(\mathcal{E}xt^1_{\mathcal{O}_{P'}}(Q_n, \mathcal{O}_{P'}^{\oplus n}))$ is a vector bundle of rank $2n^2$ on Y_n. If we set $V = \mathcal{S}pec\left(\bigoplus_{n \geq 0} S^n(F^*)\right)$, then we have the universal global section ξ of $\pi^*(F)$ on V, where $\pi : V \longrightarrow Y_n$ is the structure morphism. Each point z of $V(k)$ represents an element of $F(\pi(z)) \simeq H^0(P, \mathcal{E}xt^1_{\mathcal{O}_{P'}}(Q_n, \mathcal{O}_{P'}^{\oplus n})(\pi(z)) \simeq H^0(P, \mathcal{E}xt^1_{\mathcal{O}_P}(Q_n(\pi(z)), \mathcal{O}_P^{\oplus n})) \simeq \mathrm{Ext}^1_{\mathcal{O}_P}(Q_n(\pi(z)), \mathcal{O}_P^{\oplus n})$ by Lemma 1.4, (2) and (4). And $\xi(z)$ provides us with an element in $\pi^*(F)(z) = F(\pi(z)) \simeq \mathrm{Ext}^1_{\mathcal{O}_P}(Q_n(\pi(z)), \mathcal{O}_P^{\oplus n})$. By the above lemma we have that $H^0(V, \pi^*(F)) \simeq H^0(P_V, \mathcal{E}xt^1_{\mathcal{O}_{P_V}}((\pi\times 1)^*(Q_n), \mathcal{O}_{P_V}^{\oplus n})) \simeq \mathrm{Ext}^1_{\mathcal{O}_{P_V}}((\pi\times 1)^*(Q_n), \mathcal{O}_{P_V}^{\oplus n})$. Thus ξ defines a universal family of extensions on P

$$\mathrm{Ex} : \quad 0 \longrightarrow \mathcal{O}_{P_V}^{\oplus n} \longrightarrow E(V) \longrightarrow (\pi\times 1)^*(Q_n) \longrightarrow 0$$

By the above argument we infer that for every z in $V(k)$, the extension $\mathrm{Ex}(z)$ corresponds to the $\xi(z)$, which implies that Ex parametrizes bijectively all the extensions of $Q_n(y)$ by $\mathcal{O}_P^{\oplus n}$, where y ranges over all the elements of $Y(k)$.

§2. The moduli space $M(n, n)$.

Let us fix a k-rational point x of P. Let M be the subset of $M(n, n)$ consisting of vector bundles E which satisfy the condition of the latter half of Lemma 1.1 for the x, that is, E has exactly n jumping lines passing through x. Then M is a non-empty open subscheme of $M(n, n)$. There are a principal fibre bundle $g : \tilde{M} \longrightarrow M$ with group $PGL(N)$ and a vector bundle $\tilde{E}$ on $\tilde{M}\times P$ such that for every z in $\tilde{M}(k)$, $\tilde{E}(z)$ corresponds to the $g(z)$ in $M(k)$ and tha that $\tilde{E}$ carries a $GL(N)$-linearization with respect to the action of $GL(N)$ on $\tilde{M}\times P$ induced by that of $PGL(N)$ (see [4] §6). Note that the center of $GL(N)$ acts on $\tilde{E}$ as the multiplication by scalars. Let I be the ideal of the point x in P and q_i the i-th projection of the product $\tilde{M}\times P$. By Lemma 1.1 and the base change theorem, we see that

(2.1.1) $H = q_{1*}(\tilde{E} \otimes q_2^*(I))$ is a locally free sheaf of rank n,

(2.1.2) the natural homomorphism $\alpha : \tilde{H} = q_1^*(H) \longrightarrow \tilde{E}$ is injective,

(2.1.3) the determinant of α has the property for T in (1.3),

(2.1.4) $\tilde{R} = \tilde{E}/\tilde{H}$ is flat over $\tilde{M}$,

(2.1.5) $\tilde{R}$ has the property for R in (1.3).

We see moreover that the following exact sequence is $GL(N)$-linearized:

(2.2) $0 \longrightarrow \tilde{H} \xrightarrow{\alpha} \tilde{E} \xrightarrow{\beta} \tilde{R} \longrightarrow 0$

By the descent theory used in the proof of (1.3), it is easy to see that $\tilde{R}' = \mathcal{H}om_{\mathcal{O}_{\tilde{M}\times P}}(\tilde{R}, \tilde{R})$ is isomorphic to $\tilde{R}$ locally over $\tilde{M}$. Thus $\tilde{R}'$ is flat over $\tilde{M}$. Note that the action of $GL(N)$ on $\tilde{R}'$ is different from that on $\tilde{R}$. Indeed, the action of the center of $GL(N)$ on $\tilde{R}'$ is trivial. Thus $q_{1*}(\tilde{R}')$ descends to a coherent sheaf R_0 on M. Similarly, $q_{1*}(\mathcal{H}om(\tilde{H}, \tilde{H}))$ descends to H_0 on M, too. For every z in $\tilde{M}(k)$, $\tilde{R}'(z) \simeq \bigoplus_{i=1}^{n} \mathcal{O}_{\ell_i}$ with ℓ_i mutually distinct lines passing through x by virtue of Lemma 1.1. Thus $h^0(P, \tilde{R}'(z)) = n$ and hence $q_{1*}(\tilde{R}')$ is a locally free sheaf of rank n. Therefore, H_0 and R_0 are locally free sheaves of rank n^2 and n, respectively. Set $Z = P(H_0^* \oplus R_0^*)$ and $U = \mathcal{S}pec(\bigoplus_{m\geq 0} S^m(H_0^* \oplus R_0^*))$.

Let us prove that Z is birational to V in §1. This shows that $M(n, n)$ is stably rational as required because Y_n is rational and hence so is V. By (1.3) there are open set $\tilde{M}_0$ of $\tilde{M}$ and a morphism f of $\tilde{M}_0$ to Y_n such that $\tilde{R}_{\tilde{M}_0} \simeq (f\times 1)^*(Q_n)$ and $\tilde{H}_{\tilde{M}_0}$ is trivial. On the other hand, on $(U \times_M \tilde{M}) \times_k P$ there are universal families of endomorphisms of $\tilde{H}$ and $\tilde{R}$; $\gamma : q_{23}^*(\tilde{H}) \longrightarrow q_{23}^*(\tilde{H})$ and $\delta : q_{23}^*(\tilde{R}) \longrightarrow q_{23}^*(\tilde{R})$, where $q_{23} : (U \times_M \tilde{M}) \times_k P \longrightarrow \tilde{M} \times_k P$ is the projection. Moreover, both γ and δ are $GL(N)$-invariant. We have a non-empty open set U_0 of U such that for all y in $(U_0 \times_M \tilde{M})(k)$, $\gamma(y)$ and $\delta(y)$ are isomorphisms. Set $\tilde{U} = U_0 \times_M \tilde{M}_0$ and fix a trivialization of $\tilde{H}_{\tilde{M}_0}$ and an isomorphism of $\tilde{R}_{\tilde{M}_0}$ to $(f\times 1)^*(Q_n)$. On $P_{\tilde{U}} = \tilde{U} \times_k P$ we have an injection $\tilde{\alpha} : \mathcal{O}_{P_{\tilde{U}}}^{\oplus n} \simeq q_{23}^*(\tilde{H}) \xrightarrow{\gamma} q_{23}^*(\tilde{H}) \xrightarrow{q_{23}^*(\alpha)} q_{23}^*(\tilde{E})$ and a surjection $\tilde{\beta} : q_{23}^*(\tilde{E}) \xrightarrow{q_{23}^*(\beta)} q_{23}^*(\tilde{R}) \xrightarrow{\delta} q_{23}^*(\tilde{R}) \simeq (fp_2\times 1)^*(Q_n)$, where $p_2 : \tilde{U} \longrightarrow \tilde{M}_0$ is the projection and where we confuse the notation on $U \times_M \tilde{M}$ with its restriction to $\tilde{U}$. Thus we obtain an exact sequence on $P_{\tilde{U}}$

$$(2.3) \quad 0 \longrightarrow \mathcal{O}_{P_{\tilde{U}}}^{\oplus n} \xrightarrow{\tilde{\alpha}} q_{23}{}^*(\tilde{E}) \xrightarrow{\tilde{\beta}} (fp_2\times 1)^*(Q_n) \longrightarrow 0.$$

The GL(N)-invariance of α, β, γ and δ implies that the above exact sequence is GL(N)-invariant. This sequence defines an element $\tilde{\eta}$ of $\mathrm{Ext}^1_{\mathcal{O}_{P_{\tilde{U}}}}((fp_2\times 1)^*(Q_n), \mathcal{O}_{P_{\tilde{U}}}^{\oplus n}) \simeq H^0(P_{\tilde{U}}, \mathcal{E}xt^1_{\mathcal{O}_{P_{\tilde{U}}}}((fp_2\times 1)^*(Q_n), \mathcal{O}_{P_{\tilde{U}}}^{\oplus n})) \simeq$ $H^0(\tilde{U}, p_{12}{}^*(\mathcal{E}xt^1_{\mathcal{O}_{P_{\tilde{U}}}}((fp_2\times 1)^*(Q_n), \mathcal{O}_{P_{\tilde{U}}}^{\oplus n}))) \simeq H^0(\tilde{U}, (fp_2)^*p_{1*}(\mathcal{E}xt^1_{\mathcal{O}_{P'}}(Q_n,$ $\mathcal{O}_{P'}^{\oplus n})))$ by Lemma 1.4. Thanks to the universality of V, $\tilde{\eta}$ provides us with a morphism $\tilde{\Phi}$ of $\tilde{U}$ to V. Furthermore, GL(N)-invariance of $\tilde{\eta}$ implies that $\tilde{\Phi}$ is compatible with the GL(N)-action of on $\tilde{U}$ and the trivial action of GL(N) on V. It is easy to see that for every y in $\tilde{U}(k)$ and every λ in $k-\{0\}$, $(\tilde{\alpha}(y), \tilde{\beta}(y))$ and $(\lambda\tilde{\alpha}(y), \lambda\tilde{\beta}(y))$ give rise to the same extension. We have therefore a morphism Φ of Z_0 to V such that $\Phi\rho = \tilde{\Phi}$, where ρ is the natural morphism of $\tilde{U}$ to Z and Z_0 is the open subscheme $\rho(\tilde{U})$ of Z. For general points z of V(k), E(V)(z) is a stable vector bundle and the exact sequence

$$0 \longrightarrow \mathcal{O}_P^{\oplus n} \longrightarrow E(V)(z) \longrightarrow Q_n(\pi(z)) \longrightarrow 0$$

is unique up to isomorphisms. This means that Φ is generically surjective and hence k(V) is a subfield of k(Z) via Φ.

Since $\tilde{M}$ is an open subscheme of a Quot-scheme and $\tilde{E}$ is the universal quotient sheaf ([3]), there are a non-empty open set V_0 of V and a morphism $\tilde{h}$ of V_0 to $\tilde{M}$ such that $(\tilde{h}\times 1)^*(\tilde{E}) \simeq E(V)_{V_0}$. Moreover, $h = g\tilde{h} : V_0 \longrightarrow M$ is the morphism obtained by applying the universal property of the moduli space M to $E(V)_{V_0}$. Shrinking V_0 suitably, we may assume that $\tilde{h}(V_0) \subset \tilde{M}_0$. Then we have an exact sequence

$$\begin{array}{ccccccccc} 0 & \longrightarrow & (\tilde{h}\times 1)^*(H) & \longrightarrow & E(V)_{V_0} & \longrightarrow & (\tilde{h}\times 1)^*(\tilde{R}) & \longrightarrow & 0. \\ & & \wr\| & & & & \|\wr & & \\ & & \mathcal{O}_{P_{V_0}}^{\oplus n} & & & & (\tilde{h}f\times 1)^*(Q_n) & & \end{array}$$

It is clear that $\tilde{h}f = \pi$. On the other hand, we have another injection

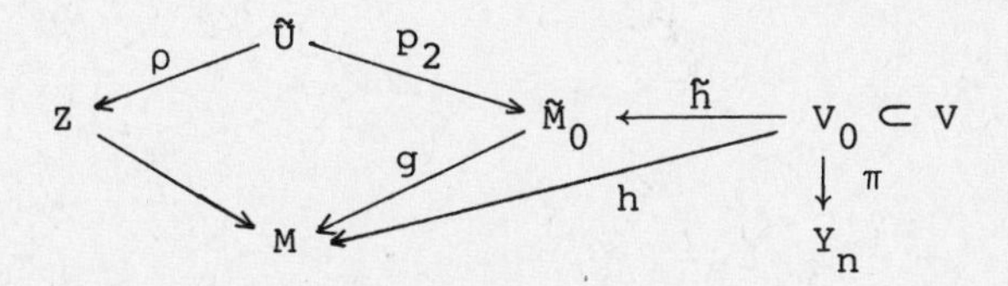

of $\mathcal{O}_{P_{V_0}}^{\oplus n}$ to $E(V)_{V_0}$ supplied by the extension Ex. Both of the trivial bundles are isomorphic to $r_1{}^*r_{1*}(E(V)_{V_0} \otimes r_2{}^*(I))$ as subsheaves of $E(V)_{V_0}$, where r_i is the i-th projection of the product $V_0 \times P$. Thus we have the following exact commutative diagram.

$$\begin{array}{ccccccccc}
 & 0 & \longrightarrow & (\tilde{h}\times 1)^*(H) & \longrightarrow & E(V)_{V_0} & \longrightarrow & (\tilde{h}\times 1)^*(\tilde{R}) & \longrightarrow & 0 \\
 & & & \wr\| & & \| & & \wr\| & & \\
 & & & \mathcal{O}_{P_{V_0}}^{\oplus n} & & \| & & (\pi\times 1)^*(Q_n) & & \\
 & & & \varepsilon\downarrow\wr & & \| & & \varepsilon'\downarrow\wr & & \\
(Ex)_{V_0} : & 0 & \longrightarrow & \mathcal{O}_{P_{V_0}}^{+n} & \longrightarrow & E(V)_{V_0} & \longrightarrow & (\pi\times 1)^*(Q_n) & \longrightarrow & 0
\end{array}$$

The couple $(\varepsilon, \varepsilon')$ yields a morphism $\tilde{\Psi} : V_0 \longrightarrow \tilde{U}$ such that $p_2\tilde{\Psi} = \tilde{h}$ and the pull back of (2.3) by $\tilde{\Psi}$ is isomorphic to $(Ex)_{V_0}$ as extensions. Therefore, we have a morphism $\Psi = \rho\tilde{\Psi}$ of V_0 to Z. By Lemma 1.1 Ψ is generically surjective and hence $k(Z)$ is a subfield of $k(V)$ via Ψ.

Since $\tilde{\Phi}$ is defined by $\tilde{\eta}$, the pull back of Ex by $\tilde{\Phi}$ is isomorphic to (2.3) as extensions. Thus the pull back of Ex by $\tilde{\Phi}\tilde{\Psi}$ is isomorphic to $(Ex)_{V_0}$ as extensions. This means that $\tilde{\Phi}\tilde{\Psi}$ is the identity on V_0. Since $\tilde{\Phi} = \Phi\rho$, we have $id_{V_0} = (\Phi\rho)\tilde{\Psi} = \Phi(\rho\tilde{\Psi}) = \Phi\Psi$ which implies that the map $k(Z) \longrightarrow k(V)$ induced by Ψ is surjective. We have proved the following theorem.

<u>Theorem 2.4</u>. <u>Let</u> $M(n, n)$ <u>be the moduli space of vector bundles</u> E <u>on the projective plane</u> P_k^2 <u>with</u> $c_1(E) = 0$ <u>and</u> $c_2(E) = r(E) = n$. <u>Then</u> $M(n, n)$ <u>is stably rational over</u> k.

Let $\nu : C(n) \longrightarrow B$ be the family of non-singular plane curves of degree n, where B is an open subscheme in P_k^M with $M = n(n+3)/2$. For the genus $g = (n-1)(n-2)/2$ of $C(n)$, we set $J(n)$ to be the connected component of the relative Picard scheme of $C(n)$ which parametrizes line bundles of degree $g-1$. We proved in [6] that $J(n)$ is birational to $M(n, n)$ ([6] Corollary 1.15.4).

<u>Corollary 2.4.1</u>. $J(n)$ <u>is stably rational</u>.

References

[1] J. M. Drezet et J. Le Potier, Fibrés stables et fibrés exceptionals, to appear.

[2] K. Hulek, On the classification of stable rank-r vector bundles over the projective plane, Proceedings, Vector bundles and differencial equations, Nice, 1979 (ed. A. Hirschowitz), Progress in Math. 7, Birkhäuser, Boston-Basel-Stuttgart.

[3] M. Maruyama, Moduli of stable sheaves, I, J. Math. Kyoto Univ., 17, 1977, 91-126.

[4] M. Maruyama, Moduli of stable sheaves, II, J. Math. Kyoto Univ., 18, 1978, 557-614.

[5] M. Maruyama, Vector bundles on P^2 and torsion sheaves on the dual plane, to appear in the Proceedings of Intern. Colloq. on Vector Bundles, Tata Inst. Fundamental Research, 1984.

[6] M. Maruyama, The equations of plane curves and the moduli spaces of vector bundles on P^2, to appear.

Department of Mathematics
Faculty of Science
Kyoto University
Sakyo-ku, Kyoto 606
Japan

Compact Kähler Manifolds of Nonnegative Holomorphic Bisectional Curvature

Ngaiming Mok

As a generalization of the elliptic and parabolic cases of the uniformization theorem in one complex variable to higher dimensions, we prove

Main Theorem

Let (X,g) be an n-dimensional compact Kähler manifold of nonnegative holomorphic bisectional curvature. Let $(\tilde{X},\tilde{g})$ be its universal covering space. Then, there exists nonnegative integers $k, N_1, \ldots, N_\ell$ and irreducible compact Hermitian symmetric spaces $M_1, \ldots, M_p$ of rank $\geqslant 2$, such that $(\tilde{X},\tilde{g})$ is isometrically biholomorphic to

$$(\mathbb{C}^k, g_0) \times (\mathbb{P}^{N_1}, \theta_1) \times \cdots \times (\mathbb{P}^{N_\ell}, \theta_\ell) \times (M_1, g_1) \times \cdots \times (M_p, g_p)$$

where g_0 denotes the Euclidean metric on $\mathbb{C}^k$; $g_1, \ldots, g_p$ are canonical metrics on $M_1, \ldots, M_p$ and θ_i, $1 \leqslant i \leqslant \ell$, is a Kähler metric on $\mathbb{P}^{N_i}$ carrying nonnegative holomorphic bisectional curvature.

In 1979, Mori [7] and Siu-Yau [9] independently proved the Frankel conjecture. By their results, one knows that an n-dimensional compact Kähler manifold of positive bisectional curvature is biholomorphic to the projective space $\mathbb{P}^n$. Shortly afterwards, Siu [8] obtained a curvature characterization for hyperquadrics. Then, in 1983, Bando [1] characterizes up to biholomorphic equivalence 3-dimensional compact Kähler manifolds of nonnegative bisectional curvature. On the other hand, Mok-Zhong [6] proved in 1984 the special case of the Main Theorem assuming that (X,g) carries constant scalar curvature. Very recently, Cao-Chow [3] showed that our Main Theorem is valid under the stronger assumption that the curvature operator is nonnegative in the dual Nakano sense.

It is well-known, using the splitting theorem of Howard-Smyth-Wu [5] that one can reduce our Main Theorem to the proof of the following special case.

Theorem 1

Let (X,g) be a compact Kähler manifold of nonnegative holomorphic bisectional curvature such that the Ricci curvature is positive at one point. Assume the second Betti number of X equals one. Then, either X is biholomorphic to the complex projective space or (X,g) is isometrically biholomorphic to an irreducible compact Hermitian symmetric spaces of rank $\geqslant 2$.

There are two distinct approaches to special cases of our Main Theorem. In the works of Mori [7], Siu-Yau [9] and Siu [8] on the analytic characterization of either the projective space or the hyperquadric, the main object under study is the space of rational curves on X. In the partial results of Bando [1], Mok-Zhong [6] and Cao-Chow [3], where the metric structure plays a more substantial role, the main technique comes from sophisticated forms of the maximum principle on tensors. The key idea in our proof is the understanding of the interplay between the two distinct aspects of the problem in the general setting.

By a classical result of Berger [2], to prove Theorem 1 it suffices to show that either X is biholomorphic to the complex projective space or there exists a proper subset S of the projectivized tangent bundle $\mathbb{P}T_X$ such that S is invariant under holonomy. Suppose M is a compact Hermitian symmetric space of rank ≥ 2. Let $\mathcal{M} \subset \mathbb{P}T_M$ be the space of tangent directions arising from unit tangent vectors of maximal holomorphic sectional curvature. Then, $\mathcal{M}$ is invariant under holonomy. While defined using some canonical metric, $\mathcal{M}$ can however be described in algebro-geometric terms completely independent of the background Kähler metric. Namely, $\mathcal{M}$ is the set of all tangent directions to rational curves of minimal degrees in M. Our task is to recover this set $\mathcal{M}$ in the abstract situation of Theorem 1. We use the parabolic Einstein equation of Hamilton [4] and the theory of rational curves of Mori [7]. Then, we will show that the $S \subset \mathbb{P}T_X$ thus constructed is in fact invariant under holonomy, completing the proof of Theorem 1 and thus the Main Theorem.

In the present note we will give a sketch of the proof of the Main Theorem. Details of the proof will appear in a forthcoming article. The author would like to thank Cao, Mori, Siu and Yau for discussions related to the research.

Table of Contents

§1 Deforming the Kähler metric by the parabolic Einstein equation

Let (X,g) be as in Theorem 1. We study the parabolic Einstein equation $\frac{\partial}{\partial t} g_{i\bar{j}} = -R_{i\bar{j}}$ on X, where $(R_{i\bar{j}})$ stands for the Ricci form. It is well-known that short time solutions $(0 \leq t < \delta$ for some $\delta > 0)$ exist and that $g(t) = (g_{i\bar{j}}(t))$ is a Kähler metric for such t. We prove

Proposition 1 (for $n = 3$ due to Bando [1])

Let (X,g) be a compact Kähler manifold of nonnegative holomorphic bisectional curvature. Then, the evolved metrics $g(t)$, $0 \leq t < \delta$ also carries nonnegative holomorphic bisectional curvature. Moreover, if (X,g) is of positive Ricci curvature at one point, then, the evolved metrics are of positive holomorphic sectional curvature and positive Ricci curvature everywhere.

Remark

Prop. 1 will allow us to construct rational curves in §2 using results of Mori [7] in algebraic geometry.

Proof

The basis of the proof is the strong maximum principle of Hamilton [4] for tensors. Let $(R_{i\bar{j}k\bar{\ell}})$ be the metric curvature tensor. We have (cf. Bando [1])

$$\frac{\partial}{\partial t} R_{i\bar{j}k\bar{\ell}} = \Delta R_{i\bar{j}k\bar{\ell}} + F(R)_{i\bar{j}k\bar{\ell}} \qquad (1)$$

where Δ denotes the Laplacian of $(X,g(t))$. In order to prove Prop. 1 by Hamilton [4] it suffices to show the "null-vector condition"

(NVC) If $u_{\chi\bar{\chi}\eta\bar{\eta}} \geqslant 0$ on X for some tensor u having the same universal symmetries as R and $\alpha, \zeta \in T_x X$ are such that $u(\alpha,\bar{\alpha};\zeta,\bar{\zeta})(x) = 0$, then $F(u)_{\alpha\bar{\alpha}\zeta\bar{\zeta}} \geqslant 0$

From standard commutation formulas we have

$$F(R)_{\alpha\bar{\alpha}\zeta\bar{\zeta}} = \sum_{\mu,\nu} R_{\alpha\bar{\alpha}\mu\bar{\nu}} R_{\nu\bar{\mu}\zeta\bar{\zeta}} - \sum_{\mu,\nu} |R_{\alpha\bar{\mu}\zeta\bar{\nu}}|^2 + \sum_{\mu,\nu} |R_{\alpha\bar{\zeta}\mu\bar{\nu}}|^2 \qquad (2)$$

whenever $R_{\alpha\bar{\alpha}\zeta\bar{\zeta}} = 0$ and $R_{\chi\bar{\chi}\eta\bar{\eta}} \geqslant 0$ on X. To prove (NVC) we are going to show

$$\text{(A)} \qquad \sum_{\mu,\nu} R_{\alpha\bar{\alpha}\mu\bar{\nu}} R_{\nu\bar{\mu}\zeta\bar{\zeta}} \geqslant \sum_{\mu,\nu} |R_{\alpha\bar{\mu}\zeta\bar{\nu}}|^2$$

Consider the Hermitian form $H_\alpha(\mu,\nu) = R_{\alpha\bar{\alpha}\mu\bar{\nu}}$ attached to α and let $\{e_\mu\}$ be an orthonormal basis of eigenvectors of H_α. In this basis we have

$$\sum_{\mu,\nu} R_{\alpha\bar{\alpha}\mu\bar{\nu}} R_{\zeta\bar{\zeta}\nu\bar{\mu}} = \sum_{\mu} R_{\alpha\bar{\alpha}\mu\bar{\mu}} R_{\mu\bar{\mu}\zeta\bar{\zeta}} . \qquad (3)$$

As in Mok-Zhong [6] we consider the function

$$G_\chi(\epsilon) = R(\alpha + \epsilon\chi, \overline{\alpha + \epsilon\chi}, \zeta + \epsilon\sum_\mu C_\mu e_\mu, \overline{\zeta + \sum_\mu C_\mu e_\mu}) \qquad (4)$$

We have $G_\chi(0) = 0$, $G_\chi(\epsilon) \geq 0$ so that $\frac{\partial^2 G_\chi}{\partial \epsilon^2}(0) \geq 0$. Thus,

$$\frac{\partial^2 G_\chi}{\partial \epsilon^2}(0) = R_{\chi\bar{\chi}\zeta\bar{\zeta}} + \sum_\mu |C_\mu|^2 R_{\alpha\bar{\alpha}\mu\bar{\mu}} + 2\mathrm{Re} \sum_\mu \bar{C}_\mu R_{\alpha\bar{\chi}\zeta\bar{\mu}}$$

$$+ 2\mathrm{Re} \sum_\mu C_\mu R_{\alpha\bar{\zeta}\mu\bar{\chi}} \geq 0 \tag{5}$$

For χ fixed and C_μ real regard $\frac{\partial^2 G_\chi}{\partial \epsilon^2}(0)$ as a positive semi-definite quadratic polynomial in C_μ. Note that (5) remains valid if e_μ is replaced by $e^{i\theta_\mu} e_\mu$. Observing this, it follows readily that if $R_{\alpha\bar{\alpha}\mu\bar{\mu}} = 0$ then (5) implies $R_{\alpha\bar{\chi}\zeta\bar{\mu}} = R_{\alpha\bar{\zeta}\mu\bar{\chi}} = 0$ for any tangent vector χ (by polarization). Order the eigenvectors of H_α so that $R_{\alpha\bar{\alpha}\mu\bar{\mu}} \neq 0$ if and only if $1 \leq \mu \leq m$. We compute the discriminant of (5) as a real quadratic polynomial in the real variables $(C_\mu)_{1 \leq \mu \leq m}$. Replacing e_μ by $e^{i\theta_\mu} e_\mu$ for appropriate choices of θ_μ we obtain

$$R_{\chi\bar{\chi}\zeta\bar{\zeta}} \geq \sum_{1 \leq \mu \leq m} \frac{1}{R_{\alpha\bar{\alpha}\mu\bar{\mu}}} | R_{\alpha\bar{\zeta}\mu\bar{\chi}} + \overline{R_{\alpha\bar{\chi}\zeta\bar{\mu}}}|^2 \tag{6}$$

Averaging over $e^{i\theta}\alpha$, $0 \leq \theta \leq 2\pi$, we obtain by taking $\chi = e_\nu$ and adding over ν, $1 \leq \nu \leq m$

$$\sum_\mu R_{\alpha\bar{\alpha}\mu\bar{\mu}} R_{\mu\bar{\mu}\zeta\bar{\zeta}} \geq \sum_{\mu,\nu} \frac{R_{\alpha\bar{\alpha}\mu\bar{\mu}}}{R_{\alpha\bar{\alpha}\nu\bar{\nu}}} (|R_{\alpha\bar{\zeta}\nu\bar{\mu}}|^2 + |R_{\alpha\bar{\nu}\zeta\bar{\mu}}|^2)$$

$$\geq \tfrac{1}{2}\sum_{\mu,\nu} \left(\frac{R_{\alpha\bar{\alpha}\mu\bar{\mu}}}{R_{\alpha\bar{\alpha}\nu\bar{\nu}}} + \frac{R_{\alpha\bar{\alpha}\nu\bar{\nu}}}{R_{\alpha\bar{\alpha}\mu\bar{\mu}}}\right) (|R_{\alpha\bar{\mu}\zeta\bar{\nu}}|^2)$$

$$\geq \sum_{\mu,\nu} |R_{\alpha\bar{\mu}\zeta\bar{\nu}}|^2 \tag{7}$$

In particular, this proves (A) and thus (NVC). It also yields immediately that if $0 < t < \delta$ and $R_{\alpha\bar{\alpha}\zeta\bar{\zeta}}(t) = 0$, we have $R_{\alpha\bar{\zeta}i\bar{j}}(t) = 0$ for all i, j. By the standard maximal principle for the heat equation, $(X,g(t))$ carries positive Ricci curvature if (X,g) carries positive Ricci curvature at one point. Suppose $R_{\alpha\bar{\alpha}\alpha\bar{\alpha}}(t) = 0$. Then $R_{\alpha\bar{\alpha}i\bar{j}}(t) = 0$ for all i,j so that $R_{\alpha\bar{\alpha}}(t) = 0$, contradicting the positivity of $(R_{i\bar{j}}(t))$. Thus, $(X,g(t))$ carries positive holomorphic sectional curvature for $t > 0$. The proof of Prop. 1 is thus completed.

§2 The space of rational curves of minimal degree

We formulate here a result of Mori [7] on rational curves.

Theorem (Mori [7])

Let X be an n-dimensional compact complex manifold with ample anti-canonical bundle K_X^{-1}. Then, there exists a rational curve $f: \mathbb{P}^1 \mapsto X$ such that $f^*(K_X^{-1}) \cong \mathcal{O}(q)$ with $0 < q \leqslant n+1$. Moreover, if there exists $P \in X$ such that $f^*T_X \cong \mathcal{O}(2) \oplus \mathcal{O}(1)^{n-1}$ for every rational curve passing through P, where T_X stands for the holomorphic tangent bundle, then X is biholomorphic to the projective space $\mathbb{P}^n$.

Remark

Mori's result is stronger than stated if we assume that X is algebraic.

From the results of §1 we may assume without loss of generality that the Kähler manifold (X,g) of Theorem 1 carries positive Ricci curvature. Then Mori's result and the Embedding Theorem of Kodaira guarantee the existence of a single rational curve $f_\circ : \mathbb{P}^1 \mapsto X$ such that $f_\circ^*(K_X^{-1}) \cong \mathcal{O}(q)$. Consider $G = \text{Graph } (f_\circ) \subset \mathbb{P}^1 \times X$. We study the deformation of G inside $\mathbb{P}^1 \times X$. Let $N_{G|\mathbb{P}^1 \times X}$ denote the normal bundle of G in $\mathbb{P}^1 \times X$. We have $H^0(G, N_{G|\mathbb{P}^1 \times X}) \cong H^0(\mathbb{P}^1, f_\circ^* T_X)$

while $H^1(G, N_{G|\mathbb{P}^1\times X}) \cong H^1(\mathbb{P}^1, f_\circ^* T_X) = 0$ because $f_\circ^* T_X \cong \sum_{i=1}^{n} \mathcal{O}(a_i)$ with $a_i \geq 0$ since (X,g) carries nonnegative holomorphic bisectional curvature. Thus the infinitesimal deformation of G inside $\mathbb{P}^1 \times X$ is parametrized by $H^0(\mathbb{P}^1, f_\circ^* T_X) \cong \mathbb{C}^{q+n}$. Let $\mathcal{D}$ be the connected component of the Douady space of $\mathbb{P}^1 \times X$ containing G and let $\mathcal{H} \subset \mathcal{D}$ be the Zariski-dense subset representing graphs of holomorphic maps $f: \mathbb{P}^1 \to X$. It is easy to see that $\mathcal{H}$ is a smooth manifold. $\mathrm{Aut}(\mathbb{P}^1)$ acts on $\mathcal{H}$ on the right by composition. The rational curves $\{f(\mathbb{P}^1) : f \in \mathcal{H}\}$ are parametrized by $\mathcal{H}/\mathrm{Aut}(\mathbb{P}^1)$. It is not difficult to show that $v = \mathcal{H}/\mathrm{Aut}(\mathbb{P}^1)$ is compact and non-singular if we take q to be the minimal possible integer. We call rational curves $f(\mathbb{P}^1)$ thus obtained rational curves of minimal degree. We are going to construct $\mathcal{S} \subset \mathbb{P}T_X$ in two different cases:

(i) $f^* T_X \cong \mathcal{O}(a_1) \oplus \cdots \oplus \mathcal{O}(a_{n-\ell}) \oplus \mathcal{O}^\ell$, $a_1 \geq \cdots \geq a_{n-\ell} > 0$
$\ell > 0$, for a generic $[f] \in \mathcal{H}$.

(ii) $q = n + 1$, $f^* T_X \cong \mathcal{O}(2) \oplus \mathcal{O}(1)^{n-1}$ for a generic $[f] \in \mathcal{H}$ but for every $x \in X$ there exists at least one $[f] \in \mathcal{H}$ with $x \in f(\mathbb{P}^1)$ such that $f^* T_X \not\cong \mathcal{O}(2) \oplus \mathcal{O}(1)^{n-1}$.

We note that since $f = \mathbb{P}^1 \mapsto X$ is a non-trivial holomorphic map under all circumstances at least one component of $f^* T_X$ must be of degree ≥ 2, so that (i) is automatically satisfied whenever $q < n+1$. When $q = n+1$, then a priori both (i) and (ii) can occur, or there exists some point $x \in X$ such that $f^* T_X \cong \mathcal{O}(2) \oplus \mathcal{O}(1)^{n-1}$ for all $[f] \in \mathcal{H}$ passing through x, in which case it follows from Mori's result that $X \cong \mathbb{P}^n$. Thus, to prove Theorem 1 we will consider cases (i) and (ii) and construct corresponding $\mathcal{S} \subset \mathbb{P}T_X$ invariant under holonomy. In case of (ii) we call $f: \mathbb{P}^1 \to X$ a special rational curve of minimal degree if $f^* T_X \not\cong \mathcal{O}(2) \oplus \mathcal{O}(1)^{n-1}$. We will construct $S_\circ \subset \mathbb{P}T_X$ and define $S = \bar{S}_\circ$. $S_\circ$ is defined in the two cases by

Case (i): $S_\circ$ is the collection of all tangent directions $[\alpha] \in \mathbb{P}T_x(X)$, $x \in X$, to rational curves C of minimal degree at regular points of C

Case (ii): as in Case (i) but one considers only special rational curves C.

We assert

Proposition 2

In case (i) $S \subset \mathbb{P}T_X$ is a subvariety of codimension $\ell \geqslant 1$. In case (ii) $S \subset \mathbb{P}T_X$ is a (possibly singular) hypersurface.

Sketch of proof

Case (i): For a generic $[f] \in H$, $f^*(T_X) \cong \mathcal{O}(a_1)\oplus\cdots\oplus \mathcal{O}(a_{n-\ell}) \oplus \mathcal{O}^\ell$. It is easy to see that $S \subset \mathbb{P}T_X$ is an irreducible subvariety. Consider a point $[\alpha] \in S_x = S \cap \mathbb{P}T_x(X)$ such that both S_x and S are smooth at x and $\mathbb{C}_\alpha = T_x(C)$ for some generic rational curve C of minimal degree smooth at x. It suffices to show that $T_{[\alpha]}(S_x)$ is of codimension ℓ in $T_{[\alpha]}(\mathbb{P}T_x(X))$. To see this one considers deformations of the rational curve C. Let $H(x)$ be the space of holomorphic maps $f\colon \mathbb{P}^1 \mapsto X$ such that $f(0) = x$. Then, in a natural way $H(x)$ is a complex space. Let $P = (0,x)$. For our rational curve C let $C = g(\mathbb{P}^1)$. $[g] \in H(x)$. Infinitesimal deformations of $G = \text{graph}(g)$ is described by $H^0(G, N_{G|\mathbb{P}^1\times X} \boxtimes \mathcal{O}(-P))$.

The obstruction is given by $H^1(G,N_{G|\mathbb{P}^1\times X} \boxtimes \mathcal{O}(-P)) \cong H^1(\mathbb{P}^1, g^*T_X \boxtimes \mathcal{O}(-1)) = 0$ since $H^1(\mathbb{P}^1, \mathcal{O}(a)) = 0$ on $\mathbb{P}^1$ if $a \geqslant -1$. Thus $H(x)$ is smooth at $[g] \in H(x)$. Let V_x be the $\mathbb{C}$-linear subspace in $T_x(X)$

corresponding to the positive components in the decomposition $g^*T_X \cong \mathcal{O}(a_1)\oplus\cdots\oplus \mathcal{O}(a_{n-\ell})\oplus \mathcal{O}^{\ell}$. Then, it is easy to check that $\alpha \in V$ and that one can identify $T_{[\alpha]}(S_x)$ with $V_x/\mathbb{C}\alpha$. Clearly, $T_{[\alpha]}(S_x)$ is of codimension ℓ in $T_{[\alpha]}(\mathbb{P}T_x(X))$.

Case (ii): Let $\mathcal{W} = \mathcal{H}(x)/\mathrm{Aut}(\mathbb{P}^1,0)$ where $\mathrm{Aut}(\mathbb{P}^1,0)$ is the isotropy subgroup at 0 of $\mathrm{Aut}(\mathbb{P}^1) \cong SL(2,\mathbb{C})$. $\mathcal{W}$ is a normal complex space of dimension $(n-1)$. Let $\Theta : \mathcal{W} \mapsto \mathbb{P}T_x(X) \cong \mathbb{P}^{n-1}$ be the meromorphic map defined by setting $\Theta(w) = [\alpha]$ if $w \in \mathcal{W}$ is represented by some $g: \mathbb{P}^1 \mapsto X$ such that $g(0) = x$, x is a smooth point of $C = g(\mathbb{P}^1)$ and $T_x(C) = \mathbb{C}$. It can actually be proved that $\mathcal{W}$ is non-singular and that a generic special rational curve is smooth at x. Moreover, one can show that Θ is of rank $(n-1)$ and the closure of the branching locus corresponds to the set of special rational curves. Thus, $S_x = S \cap \mathbb{P}T_x(X)$ is of codimension one if for a generic $x \in X$, Θ is of maximal rank when restricted to the set of ξ of special rational curves C. Otherwise for a generic $x \in X$ the level sets of $\Theta|_{\xi}$ are of positive dimension. We deduce from this fact that there exists a non-trivial family of special rational curves parametrized by a non-singular algebraic curve Γ passing through two distinct points $x,y \in X$. In other words, there exists a holomorphic $\mathbb{P}^1$ bundle over Γ and two distinct holomorphic sections which are blown down to x and y respectively. This contradicts easily the criterion of Mumford-Grauert on blowing down curves in surfaces.

§3 Invariance of S under holonomy

$S \subset \mathbb{P}T_X$ is an irreducible complex subvariety. To prove that S is invariant under holonomy it suffices to show that S is in some sense infinitesimally invariant under parallel transport. Recall that in case (i), for a generic $[\alpha] \in S_x$, $T_{[\alpha]}(S_x) \cong V_x/\mathbb{C}\alpha$ where

$V_x \subset T_x(\mathbb{C})$ corresponds to the positive components in the decomposition $f^*T_X \cong \mathcal{O}(a_1) \oplus \cdots \oplus \mathcal{O}(a_{n-\ell}) \oplus \mathcal{O}^{\ell}$, where $f\colon \mathbb{P}^1 \mapsto X$ represents a rational curve with $f(0) = x$ and $T(f(\mathbb{P}^1)) = \mathbb{C}\alpha$. A similar situation holds for Case (ii). To prove the invariance of S under holonomy in both cases it suffices to show

Proposition 3

Let $[\alpha] \in S_x$ be a smooth point of both S_x and S. Let $\gamma\colon (-\delta,\delta) \mapsto X$ be a geodesic interval such that $\gamma(0) = x$ and $\dot{\gamma}(0) = \eta$. Let now $\alpha(t) \in T_{\gamma(t)}(X)$ such that $[\alpha(t)] \in S_{\gamma(t)}$ and $\alpha\colon (-\delta,\delta) \mapsto T_X$ is smooth. Then $(\nabla_\eta \alpha)(0) \in V_x$.

Here we have chosen a Kähler metric g on X obtained by evolution. Thus, (X,g) carries nonnegative holomorphic bisectional curvature, positive holomorphic sectional curvature and positive Ricci curvature. ∇ will denote covariant differentiation with respect to (X,g).

Proof of Prop. 3

Let $\beta\colon (-\delta,\delta) \to T_X$, $\beta(t) \in T_{\gamma(t)}$ be obtained from $\alpha(0) = \alpha$ by parallel transport. i.e. $\beta(0) = \alpha$ and $\nabla_{\gamma(t)} \beta \equiv 0$. At each $\gamma(t)$, let $\mathcal{N}_{\gamma(t)}$ be the null-space of the Hermitian bilinear form $H_{\alpha(t)}(\xi,\eta) = R_{\alpha(t)\overline{\alpha(t)}\xi\overline{\eta}}$. Let $\mathcal{D}_{\gamma(t)}$ be the orthogonal complement of $\mathcal{N}_{\gamma(t)} + \mathbb{C}\alpha$. Prop. 3 is proved from the statements

$$\text{(I)} \quad (\nabla_\eta \alpha)(0) \in \mathbb{C}\alpha + \mathcal{D}_{\gamma(0)} = \mathcal{N}^{\perp}_{\gamma(0)}$$

$$\text{(II)} \quad \mathcal{N}^{\perp}_{\gamma(0)} \subset V_{\gamma(0)}$$

Write

$$\beta(t) = \alpha(t) + t\xi(t) + t\zeta(t) \qquad \xi(t) \in \mathcal{D}_{\gamma(t)},\ \zeta(t) \in \mathcal{N}_{\gamma(t)}$$

Since

$$0 = (\nabla_\eta \beta)(0) = (\nabla_\eta \alpha)(0) + \xi + \zeta, \quad \xi = \xi(0), \zeta = \zeta(0).$$

to prove (I) it suffices to show that $\zeta = 0$. Consider the vector field $X(t)$ on γ obtained by parallel transport of ζ, i.e. $X(0) = \zeta$ and $\nabla_{\gamma(t)} X \equiv 0$. Write

$$X(t) = \zeta'(t) + t\eta(t), \quad \zeta'(t) \in N_{\gamma(t)}, \ \eta(t) \in N^{\perp}_{\gamma(0)}$$

Then, consider the expansion

$$R_{\beta(t)\overline{\beta(t)}X(t)\overline{X(t)}} = t^2 A + 0(t^3). \tag{8}$$

Using results from §1, we have

$$A = (R_{\alpha\bar{\alpha}\eta\bar{\eta}} + R_{\xi\bar{\xi}\zeta\bar{\zeta}} + 2\mathrm{Re}\, R_{\alpha\bar{\eta}\zeta\bar{\xi}}) + R_{\zeta\bar{\zeta}\zeta\bar{\zeta}} + 2\mathrm{Re}R_{\xi\bar{\zeta}\zeta\bar{\zeta}}$$

$$= B + R_{\zeta\bar{\zeta}\zeta\bar{\zeta}} + 2\mathrm{Re}R_{\xi\bar{\zeta}\zeta\bar{\zeta}} \tag{9}$$

Again, from consideration is §1 (eqt. (5)) we have $B \geqslant 0$. Thus, we have

$$A \geqslant R_{\zeta\bar{\zeta}\zeta\bar{\zeta}} + 2\mathrm{Re}R_{\xi\bar{\zeta}\zeta\bar{\zeta}} \tag{10}$$

On the other hand

$$\nabla^2_\eta R_{\alpha\bar{\alpha}\zeta\bar{\zeta}} = \frac{d}{dt^2} R_{\beta(t)\overline{\beta(t)}X(t)\overline{X(t)}} = A \geqslant R_{\zeta\bar{\zeta}\zeta\bar{\zeta}} + 2\mathrm{Re}R_{\xi\bar{\zeta}\zeta\bar{\zeta}} \tag{11}$$

$$0 \leqslant \Delta R_{\alpha\bar{\alpha}\zeta\bar{\zeta}} = \frac{\partial}{\partial t} R_{\alpha\bar{\alpha}\zeta\bar{\zeta}} - F(R)_{\alpha\bar{\alpha}\zeta\bar{\zeta}} \tag{12}$$

Our Kähler metric g is obtained by evolving some Kähler metric on X of nonnegative holomorphic bisectional curvature. From the stability of the nonnegativity of bisectional curvatures it follows readily that $\frac{\partial}{\partial t} R_{\alpha\bar{\alpha}\zeta\bar{\zeta}} = 0$. By §1, $F(R)_{\alpha\bar{\alpha}\zeta\bar{\zeta}} \geq 0$. This forces $\Delta R_{\alpha\bar{\alpha}\zeta\bar{\zeta}} = 0$ so that also $\nabla_\eta^2 R_{\alpha\bar{\alpha}\zeta\bar{\zeta}} = 0$. It follows then $R_{\zeta\bar{\zeta}\zeta\bar{\zeta}} + 2\mathrm{Re} R_{\xi\bar{\zeta}\zeta\bar{\zeta}} = 0$. We assert that $R_{\xi\bar{\zeta}\zeta\bar{\zeta}} = 0$. Consider the inequalities (6) and (7) from which we obtained the inequality $F(R)_{\alpha\bar{\alpha}\zeta\bar{\zeta}} \geq 0$. The equality $F(R)_{\alpha\bar{\alpha}\zeta\bar{\zeta}} = 0$ then forces (6) to be an equality. Consider the expression $g(X)$ obtained by computing discriminants of the quadratic form $\frac{\partial^2 G}{\partial \varepsilon^2}(0)$ in (5). Fixing ζ we have thus $g(X) \geq 0$ and $g(\mu_i) = 0$ $1 \leq i \leq m$ where $\{\mu_i\}$ $1 \leq i \leq n$ is a basis of $T_x(X)$ of eigenvectors of H_α and $R_{\alpha\bar{\alpha}\mu_i\bar{\mu}_i} \neq 0$ if and only if $1 \leq \mu_i \leq m$. Considering the variation equalities $\frac{\partial}{\partial \varepsilon} g(\mu_i + \varepsilon e^{i\theta}\zeta) = 0$ we obtain easily $R_{\mu_i\bar{\zeta}\zeta\bar{\zeta}} = 0$ for $1 \leq i \leq m$. Since $\xi = \sum_{i=1}^{m} a_i\mu_i$ we deduce $R_{\xi\bar{\zeta}\zeta\bar{\zeta}} = 0$. Thus we have $R_{\zeta\bar{\zeta}\zeta\bar{\zeta}} = 0$ from equations (11) and (12). Since (X,g) carries positive holomorphic sectional curvature this implies $\zeta = 0$, proving (I).

$R_{\zeta\bar{\zeta}\zeta\bar{\zeta}} = 0$. But since (X,g) carries positive holomorphic sectional curvature this implies $\zeta = 0$, proving (I).

To prove (II) we compute holomorphic bisectional curvature using the bundle decomposition $f^*T_X = (\sum_{i=1}^{n-\ell} \mathcal{O}(a_i)) \oplus \mathcal{O}^\ell$. We show that, for Θ the curvature of the dual cotangent bundle, $\Theta_{\nu^*\bar{\nu}^*\alpha\bar{\alpha}} = 0$ $\nu \in T^*{}_{\gamma(0)}$ whenever $\nu^*(V_{\gamma(0)}) = 0$. In other words, $R_{\alpha\bar{\alpha}\nu\bar{\nu}} = 0$ whenever $\nu \perp V_{\gamma(0)}$, i.e. $N_{\gamma(0)} \subset V_{\gamma(0)}$.

The proof of the Main Theorem is complete.

References

[1] Bando, S. On three dimensional compact Kähler manifolds of nonnegative bisectional curvature, J. of Diff. Geom. 19 (1984) 283-297.

[2] Berger, M. Sur les groupes d'holonomie homogène des variétés à connexion affine et des variétés riemanniennes. Bull Soc. Math. France, 83 (1955), 279-330.

[3] Cao, H.-D. and Chow, B. Compact Kähler manifolds with nonnegative curvature operator. Preprint.

[4] Hamilton, R.S. Three-manifolds with positive Ricci curvature, J. Diff. Geom. 17(1982), 255-306.

[5] Howard, A., Smyth B. and H. Wu, On compact Kähler manifolds of nonnegative bisectional curvature I, Acta Math 147(1981), 51-56.

[6] Mok, N. and Zhong, J.-Q. Curvature characterization of compact Hermitian symmetric spaces, J. Diff. Geom.

[7] Mori, S. Projective manifolds with ample tangent bundles, Ann. of Math. 110(1979), 593-606.

[8] Siu, Y.-T. Curvature characterization of hyperquadrics, Duke Math. J. 47(1980), 641-654.

[9] Siu, Y.-T. and Yau, S.-T. Compact Kähler manifolds of positive bisectional curvature, Invent. Math. 59 (1980), 189-204.

CONCAVITY, CONVEXITY AND COMPLEMENTS IN COMPLEX SPACES

Christian Okonek
Sonderforschungsbereich 170: Geometrie und Analysis
Universität Göttingen
Bunsenstraße 3-5
D-3400 Göttingen

CONTENTS

0. INTRODUCTION

Let $A \subset X$ be a closed complex subspace in a compact complex space X, defined by the ideal $I \subset \mathcal{O}_X$. The linear fibre space $\pi : N := \mathrm{Lin}(I/_{I^2}) \rightarrow A$ is the generalized normal bundle N of A in X.

In this paper we consider the following problem:

Find good sufficient conditions on N such that the complement $X \backslash A$ of A in X is q-convex in the sense of Andreotti-Grauert [1].

There are several papers dealing with this question. We only mention a few, which are closely related to ours. In [18] Schneider showed, that $X \backslash A$ is (q+k)-convex, if X is smooth and $A \subset X$ a submanifold of codimension q with metrically (k+1)-concave normal bundle. In [5] Fritzsche extended this result to arbitrary closed complex subspaces A in complex manifolds: If the normal bundle N of A in X is (k+1)-positive, the complement $X\setminus A$ is (q+k)-convex, where $q := \sup_{x \in A} \dim_{\mathbb{C}} N(x)$. Finally in [6] he showed, that $X\setminus A$ is q-convex if A is a closed subspace in a compact complex space X, such that the normal bundle is Finsler-positive. The conditions on the normal bundle in all these results were differential-geometric conditions. In [20] Sommese introduced his concept of k-ampleness, an algebro-geometric notion which coincides with ampleness in the sense of Hartshorne [12] for k = 0. In [24] Sommese showed, that the dual bundle of a globally generated k-ample vector bundle on compact manifold is (k+1)-convex. This result together with his concavity theorem [21] and the theorem of Fritzsche mentioned above allowed him to prove the following [24]:

Let $A \subset X$ be a closed submanifold of codimension q in a compact manifold X. If the normal bundle of A in X is globally generated and k-ample, $X\setminus A$ is (q+k)-convex.

The purpose of this paper is to find common generalizations of these theorems. In order to achieve this we introduce the concept of Finsler-k-positivity for linear fibre space. This is a differential-geometric notion, which coincides with Finsler-positivity for k = 1. For vector bundles of rank r this is equivalent to what Sommese calls strictly (r+k)-concave.

Here are the main results of our paper:

THEOREM: (Complements in Complex Spaces) Let $A \subset X$ be a closed subspace in a compact complex space X with Finsler-(k+1)-positive normal bundle N. Then the complement $X\setminus A$ is (R_N+k)-convex where

$R_N = \sup_{x \in A} \dim_{\mathbb{C}} N(x)$.

This theorem generalizes results of Fritzsche [5], [6] and Schneider [18].

A criterion for N to be Finsler-(k+1)-positive is the next theorem, which is just a reformulation of Sommese's concavity theorem [21]:

THEOREM: (Concavity) Let $\pi : E \to X$ be a globally generated vector bundle of rank r on a compact complex space X, such that E^* is (k+1)-convex. Then $\pi : E \to X$ is Finsler-(k+1)-positive.

The k-convexity of E^* is a differential-geometric notion. As in the smooth case ([14]) we can relate it to an algebro-geometric notion:

THEOREM: (Convexity) Let F be a globally genrated k-ample, coherent sheaf on a compact complex space X. Then the linear fibre space $\pi : \mathrm{Lin}(F) \to X$ associated to F is (k+1)-convex.

The proof of the convexity theorem goes along the lines of Sommese's proof in the smooth case. The main point is an extension of his perturbation lemma [24] to holomorphic mappings of arbitrary complex spaces. We show

THEOREM: (Perturbation) Let $f : X \to Y$ be a holomorphic map from a compact complex space X to a complex space Y. Let $q: T_Y \times_Y T_Y \to \mathbb{C}$ be a hermitean form which is positive semidefinite and has at most q non-positive eigenvalues in each point $y \in Y$. Then there exists a differentiable function φ on X such that $\delta\bar{\delta}\varphi + f^*g$ has at most k+q non-positive eigenvalues in each point $x \in X$, where $k = \max_{x \in X} \dim X_{f(x)}$.

The convexity theorem then essentially follows by an application of this theorem to the canonical map $\mathbb{P}(\mathrm{Lin}(F)) \to \mathbb{P}(H^o(X,F)^*)$.

I would like to mention, that my original motivation for this work was to extend the Lefschetz-type theorems of Sommese/Van de Ven [25] to the singular case. As in this paper, a proof of a generalized Lefschetz-theorem should have 2 parts, a differential geometric part and a topological part. In this sense our paper is the first step towards a generalization of the Lefschetz-type theorems to singular spaces.

I want to thank A. J. Sommese for suggesting this problem and helpful remarks. Also discussions with K. Fritzsche have been very useful. It is only fair to say, that almost all ideas are already contained in their papers.

1. CONCAVITY

Let us first recall some notations [4], [5], [6], [26]. Every complex space X (paracompact, Hausdorff, of finite dimension) with structure sheaf $\mathcal{O}_X$ can be endowed canonically with a fine sheaf $\mathcal{D}_X$ such that the following holds: If $f : B \to \mathbb{C}^m$ is a holomorphic map from a domain $B \subset \mathbb{C}^n$, such that $X = f^{-1}(0)$ is a 'local model', then

$$\mathcal{D}_X = C^\infty_B/_{(\mathrm{Re}\, f_1 \ldots, \mathrm{Re}\, f_m, \mathrm{Im}\, f_1 \ldots, \mathrm{Im}\, f_m)} .$$

In this way we have to every complex space an associated <u>differentiable space</u> $(X, \mathcal{D}_X)$. A <u>differentiable</u> <u>function</u> on $U \subset X$ is then defined as a section in $\mathcal{D}_X$ over U.

Using this concept of differentiable functions, it is not hard, to extend other differentiable notions to complex spaces.

We call a function $s \in \mathcal{D}_X(U)$ <u>q-convex</u> in $x \in U$, if the Leviform

$$L_s(x) : T_X(x) \otimes \overline{T_X(x)} \to \mathbb{C}$$

has at most q-1 non-positive eigenvalues on the tangent space $T_X(x)$ of X at x.

Let $\pi : E \to X$ be a <u>linear space</u> with <u>scalar product</u>

$$H : E \times_X E \to \mathbb{C} .$$

H induces a differentiable map

$$h : E \to \mathbb{R}$$

such that

$$h(z \cdot e) = |z|^2 h(e)$$

for $z \in \mathbb{C}^*$, $e \in E$.

For $x \in X$, $e \in E(x)$ we have a canonical exact sequence of vector-spaces

$$0 \to E(x) \to T_E(e) \xrightarrow{T_\pi(e)} T_X(x)$$

The <u>isolation</u> <u>number</u> $i_E(e)$ of E in e is the codimension in $T_X(x)$ of the normal space of π [6],

$$i_E(e) = \dim_{\mathbb{C}} T_X(x)/_{\mathrm{Im}\, T_\pi(e)} \cdot$$

Let $\varphi : E_U \longrightarrow U \times \mathbb{C}^r$ be a pseudotrivialization of E near x which is normal in x with respect to H [5]. Then we get an induced splitting

$$T_E(e) \cong E(x) \oplus B_E(e)$$

(depending on φ) which is orthogonal with respect to the Leviform $L_h(e)$

of h in e [5]. (E,H) is called <u>r-positive</u> in $x \in X$, if for each pseudo-trivialization of E near x which is normal in x with respect to H, the Levi-form $L_h(e)$ has at most r-1 non-negative eigenvalues on $B_E(e)$ [5].

We will need another positivity notion.

<u>DEFINITION 1:</u> Let $\pi : E \to X$ be a linear fibre space over X. A <u>Finsler-metric</u> on E is a differentiable function

$$\rho : E\setminus X \to (0,\infty)$$

which for every $z \in \mathbb{C}^*$, e E\X satisfies the equation

$$\rho(z\cdot e) = |z|^2\rho(e).$$

A Finsler-metric ρ is Finsler-k-positive in $x \in X$, if it has the following properties:

i) $\rho|_{E(x)\setminus\{0\}}$ is strictly plurisubharmonic

ii) the Levi-form $L_\rho(e)$ has at least emb $\dim_x X - i_E(e) - k+1$ negative eigenvalues for every $e \in E(x)\setminus\{0\}$.

The linear space is <u>Finsler-k-positive</u>, if it carries a Finsler-metric, which is Finsler-k-positive in every $x \in X$.

<u>REMARK 2:</u>

i) In [6] Fritzsche defined Finsler-positive linear spaces. These are precisely the Finsler-1-positive linear spaces in our sense.

ii) Let ρ be a Finsler-metric on a line bundle. Then there exists a scalar product H with $\rho = h$[6].

We will need the following facts about Finsler-k-positive linear spaces.

<u>PROPOSITION 3:</u> Let X be a complex space, $\pi : E \to X$ a linear fibre space over X.

i) if E is k-positive, E is Finsler-k-positive.

ii) if E is a line bundle, it is k-positive if and only if it is Finsler-k-positive.

iii) if E is Finsler-k-positive, $Y \subset X$ a closed subspace, then the restriction $E \times_X Y$ is Finsler-k-positive.

PROOF:

i) Let H be a scalar product which makes E k-positive. H induces a Finsler-metric h on E by $h(e) = H(e,e)$ such that the restriction of h to $E(x)\setminus\{0\}$ is strictly plurisubharmonic. Since $L_h(e)$ has at most k-1 non-negative eigenvalues on $B_E(e)$ for every pseudo-trivialization of E in x that is normal in x with respect to H, $L_h(e)$ must have at least $\dim_{\mathbb{C}} B_E(e) - k+1$ negative eigenvalues. But $\dim_{\mathbb{C}} B_E(e) = \text{emb}\dim_x X - i_E(e)$, so h makes E Finsler-k-positive.

ii) Let $\pi : L \to X$ be a line bundle with a Finsler-metric ρ which makes it Finsler-k-positive. There exists a scalar product H such that $\rho(l) = H(l,l)$. Choose a trivialization of L in $x \in X$ that is normal in x with respect to H. Then we get a decomposition $T_L(l) \cong L(x) \oplus B_L(l)$ which is orthogonal with respect to the Levi-form of ρ in $l \in L(x)\setminus\{0\}$. Since $L_\rho(l)$ is positive on $L(x)$ and has at least $\text{emb}\dim_x X - k+1$ negative eigenvalues, the restriction to $B_L(l)$ can have at most k-1 non-negative eigenvalues on $B_L(l)$.

iii) Let ρ be a Finsler-metric on E which makes $\pi : E \to Y$ Finsler-k-positive. Restricting ρ we obtain a Finsler-metric ρ_Y on $E \times_X Y$, which is restricted to $E(y)\setminus\{0\}$ strictly plurisubharmonic for each $y \in Y$. Since the Levi-form $L_{\rho_Y}(e)$ of ρ_Y in $e \in E \times_X Y\setminus Y$ is the restriction of the Levi-form of ρ in $e \in E$, $L_{\rho_Y}(e)$ can have at most $\dim_{\mathbb{C}} E(\pi(e)) + k-1$ non-negative eigenvalues. This shows, that $L_{\rho_Y}(e)$ has at least $\dim_{\mathbb{C}} T_{E\times_X Y}(y) - \dim_{\mathbb{C}} E(x) - k+1 = \text{emb}\dim_y Y - i_{E\times_X Y}(e) - k+1$ negative eigenvalues if $\pi(e) = y$.

For every linear fibre space $\pi : E \to X$ we have the associated projective fibre space [4]

$$p : \mathbb{P}(E) \to X.$$

$\mathbb{P}(E)$ carries the tautological line bundle $L(E)$.

$$L(E) \subset E \times_X \mathbb{P}(E)$$

is a linear subspace of $p^*E := E \times_X \mathbb{P}(E)$ and the inclusion determines a biholomorphic map

$$L(E) \setminus \mathbb{P}(E) \xrightarrow[\alpha]{} E \setminus X.$$

Define for each linear fibre space $\pi : E \to X$

$$r_E = \inf_{x \in X} \dim_{\mathbb{C}} E(x), \; R_E = \sup_{x \in X} \dim_{\mathbb{C}} E(x).$$

<u>PROPOSITION 4:</u> Let $\pi : E \to X$ be a Finsler-(k+1)-positive linear fibre space over X. Then the tautological line bundle $L(E)$ on $\mathbb{P}(E)$ is $(k+R_E)$-positive.

<u>PROOF:</u> Let $\tilde{\rho}$ make $\pi : E \to X$ Finsler-(k+1)-positive. We have the canonical biholomorphic map

$$\alpha : L(E) \setminus \mathbb{P}(E) \xrightarrow{\cong} E \setminus X.$$

It induces a Finsler-metric $\rho = \alpha^* \tilde{\rho}$ on the tautological line bundle over $\mathbb{P}(E)$. There exists a scalar product H with $\rho(l) = H(l,l)$. Choose a trivialisation of $L(E)$ around $\langle e \rangle \in \mathbb{P}(E)$, which is normal in $\langle e \rangle$ with respect to H. Then $T_{L(E)}(e)$ decomposes as

$$T_{L(E)}(e) \cong L(E)(\langle e \rangle) \oplus B_{L(E)}(e)$$

and this is orthogonal with respect to the Levi-form $L_\rho(e)$ of ρ in e. $L_\rho(e)$ is positive on $L(E)(\langle e \rangle)$ and has at least $\dim_{\mathbb{C}} T_E(e) - \dim_{\mathbb{C}} E(\pi(e)) - k$ negative eigenvalues, so it has at most

$$\dim_{\mathbb{C}} B_{L(E)}(e) - (\dim_{\mathbb{C}} T_E(e) - \dim_{\mathbb{C}} E(\pi(e)) - k)$$

non-negative eigenvalues on $B_{L(E)}(e)$. From

$$\dim_{\mathbb{C}} B_{L(E)}(e) - \dim_{\mathbb{C}} T_E(e) + \dim_{\mathbb{C}} E(\pi(e)) + k$$

$$= \dim_{\mathbb{C}} E(\pi(e)) + k-1.$$

we see, that $(L(E),H)$ is $(\dim_{\mathbb{C}} E(\pi(e)) + k)$-positive in $\langle e \rangle$. So H makes $L(E)$ (R_E+k)-positive.

<u>DEFINITION 5:</u> A linear fibre space $\pi : E \to X$ is called <u>q-convex</u> (<u>q-concave</u>), if there exists a Finsler-metric ρ on E such that $\rho(-\rho)$ is a q-convex in each point $e \in E\backslash X$. If in addition the restriction $\rho|_{E(x)\setminus\{0\}}$ is strictly plurisubharmonic, E is called <u>strongly q-convex</u> (<u>strongly q-concave</u>).

<u>REMARK 6:</u> In the case of a vector bundle $\pi : E \to X$ this definition is due to Sommese [21].

<u>LEMMA 7:</u> Let $\pi : E \to X$ be a linear fibre space.

i) If $\pi : E \to X$ is Finsler-k-positive, it is strongly (R_E+k)-concave.

ii) If $\pi : E \to X$ is strongly-(r_E+k)-concave, it is Finsler-k-positive.

PROOF: The linear fibre space $\pi : E \to X$ is strongly (r+k)-concave if and only if there exists a Finsler-metric ρ, strictly plurisubharmonic on $E(x)\setminus\{0\}$ for each $x \in X$, such that the Levi-form $L_\rho(e)$ has at most r+k-1 non-negative eigenvalues for each $e \in E\setminus X$. This is true if and only if $L_\rho(e)$ has at least $(\dim_{\mathbb{C}} T_E(e)-r) - k+1$ negative eigenvalues. From

$$\dim_{\mathbb{C}} T_E(e) = \dim T_X(x) - i_E(e) + \dim_{\mathbb{C}} E(x)$$

we get

$$\text{emb}\dim_X - i_E(e) + r_E \leq \dim_{\mathbb{C}} T_E(e) \leq \text{emb}\dim_x X - i_E(e) + R_E,$$

the lemmas is proved.

Using the main results of [21] and lemma 7 above we have

THEOREM 8: Let $\pi : E \to X$ be a holomorphic vector bundle on a compact complex space. If E is globally generated and E^* (k+1)-convex, E is Finsler-(k+1)-positive.

PROOF: Sommese shows, that under these conditions E is strongly (r+k+1)-concave. Now use lemma 7.

COROLLARY 9: Let $\pi : E \to X$ be a globally generated holomorphic vector bundle of rank r on a compact complex space. If E^* is (k+1)-convex, L(E) is Finsler-(r+k)-positive on $\mathbb{P}(E)$.

PROOF: This follows from proposition 4 and theorem 8.

2. CONVEXITY

Recall, that a line bundle $\pi : L \to X$ on a complex space X is k-ample, if some positive power $L^{\otimes s}$ is globally generated by finitely many sections, and the associated map

$$\varphi_{L^{\otimes s}} : X \to \mathbb{P}$$

has only fibres of dimension $\leq k$ [20].

DEFINITION 10: A coherent sheaf F on a complex space is k-ample, if the line bundle $L(\mathrm{Lin}\ F)^*$ on $\mathbb{P}(\mathrm{Lin}\ F)$ is k-ample.

REMARK 11: If F is locally free, $\pi : E = \mathrm{Lin}(F^*) \to X$ the associated vector bundle, then F is k-ample in the sense of Sommese [20].

LEMMA 12: Let F be a k-ample coherent sheaf on X, $Y \subset X$ a closed complex subspace. Then $F \otimes \mathcal{O}_Y$ is k-ample on Y.

PROOF: $\mathbb{P}(\mathrm{LIN}(F \otimes \mathcal{O}_Y)) = \mathbb{P}(\mathrm{Lin}(F)) \times_X Y$ is a closed complex subspace of $\mathbb{P}(\mathrm{Lin}(F))$ and $L(\mathrm{Lin}(F \otimes \mathcal{O}_Y))$ is the restriction of $L(\mathrm{Lin}(F))$ to this subspace. Now the lemma follows easily.

LEMMA 15: Let F be a coherent sheaf, globally generated by finitely many sections. F is k-ample on X if and only if $F \otimes \mathcal{O}_{X_{red}}$ is k-ample on X_{red}.

PROOF: $L(\mathrm{Lin}(F))^*$ is a vector bundle on $\mathbb{P}(\mathrm{Lin}(F))$ which is globally generated by finitely many sections since F is. The restriction f_{red} of the associated morphism

$$f : \mathbb{P}(\mathrm{Lin}(F)) \to \mathbb{P}$$

to $\mathbb{P}(\mathrm{Lin}(F))_{red}$ is the morphism associated to $L(\mathrm{Lin}(F \otimes \mathcal{O}_{X_{red}})) =$ $= L(\mathrm{Lin}(F))^* \times_{\mathbb{P}(\mathrm{Lin}(F))} \mathbb{P}(\mathrm{Lin}(F))_{red}$. So we have $\dim f^{-1}f(p) =$ $= \dim f_{red}^{-1} f_{red}(p)$ for each point $p \in \mathbb{P}(\mathrm{Lin}(F))$. This proves the lemma.

Now we come to the main result of this section.

THEOREM 14: Let $f : X \to Y$ be a holomorphic map from a compact complex space X to a complex space Y with Y_{red} smooth, such that $\dim X_{f(x)} \leq k$ for each $x \in X$. Let $g : T_Y \times_Y T_Y \to \mathbb{C}$ be a hermitean form on the tangent space of Y, which is positive semidefinite with at most q non-positive eigenvalues. Then there exists a non-negative differentiable function φ on X, such that $L_\varphi + f^*g$ has at most $q+k$ non-positive eigenvalues in each point $x \in X$.

For the proof we will need the following proposition.

PROPOSITION 15: Let $j : A \hookrightarrow X$ be a closed complex subspace in a (paracompact) complex space X. Assume, that we have a hermitean form H on the tangent space of X and a non-negative differentiable function φ on A such that $L_\varphi + j^*H$ has at most q non-positive eigenvalues on $T_A(x)$ for each point $x \in A$. Then there exists a non-negative extension Φ of φ to X, such that $L_\Phi + H$ has at most q non-positive eigenvalues on $T_X(x)$ for each x in an open neighbourhood of $j(A)$.

As a corollary we obtain a result of Fritzsche [5].

COROLLARY 16: Let $j : A \hookrightarrow X$ be a closed complex subspace in a complex space X, φ a q-convex function on A. There exists a q-convex function Φ on a neighbourhood of $j(A)$ which restricts to φ.

PROOF OF THE PROPOSITION: Since the proof is very similar to the proof of Fritzsche's result, we only give the idea.

Take a locally finite open cover $(U_\alpha)_{\alpha \in A}$ of X with $\bar{U}_\alpha$ compact, such that $U_\alpha \cap A$ is of the form $U_\alpha \cap A = g_\alpha^{-1}(0)$ for a holomorphic map $g_\alpha : U_\alpha \to \mathbb{C}^{q_\alpha}$. Let $(\rho_\alpha)_{\alpha \in A}$ be a partition of unity subordinate to $(U_\alpha)_{\alpha \in A}$. For an arbitrary extension $\varphi*$ of φ to [5] and $k = (k_\alpha)_{\alpha \in A}, k_\alpha \in \mathbb{N}$ let

$$\Phi_k := \varphi* + \sum_{\alpha \in A} k_\alpha \rho_\alpha \left(\sum_{i=1}^{q_\alpha} |g_{\alpha,i}|^2 \right).$$

Then for a suitable k $\Phi := \Phi_k$ has the required property.

PROOF OF THE THEOREM: Proposition 15 shows, that for the proof we can assume that X and Y are reduced. In fact, one has a commutative diagram

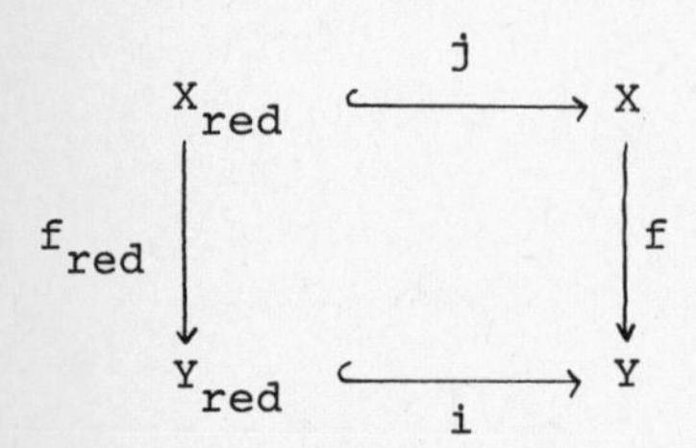

If φ is a differentiable function on X_{red}, such that $L_\varphi + (i \circ f_{red})^* g$ has at most q+k non-positive eigenvalues, proposition 15 gives an extension Φ of φ to X, such that $L_\Phi + f^* g$ has at most q+k non-positive eigenvalues.

Then we have the singular filtration of X with respect to f. This is a finite, decreasing family $X = X_o \supset X_1 \supset \ldots \supset X_m \supset \emptyset$ of closed analytic subsets, such that $X_i \setminus X_{i+1}$ is smooth and $f|_{X_i \setminus X_{i+1}}$ locally a product (cf. [24]). For every X_i one can choose a non-negative differentiable function ψ_{X_i} on X such that for $x \in X_i$ the Levi-form $L_{\psi_i}(x)$ is positive semi-definite on $T_X(x)$ and positive definite on every complement of $T_{X_i}(x)$ in $T_X(x)$.

Now consider the property (p_j):

(p_j) There exists a non-negative differentiable function φ_j on X such that for each $x \in X_j$ $L_{\varphi_j}(x) + f^*g$ is positive definite on a subspace of codimension $\leq q+k$ in $T_X(x)$

If we can prove (p_o), the theorem follows. We do induction on j.

For every j we have a function ψ_{X_j} on X such that $L_{\psi_{X_j}} + f^*g$ is positive semidefinite on $T_X(x)$ for $x \in X_j$ and positive definite on a subspace of codimension $\leq q+k$ in $T_X(x)$ if $x \in X_j \setminus X_{j+1}$. The second property is true, because $f|_{X_j \setminus X_{j+1}}$ is locally a product on $X_j \setminus X_{j+1}$. Taking $j = m$ we get property (p_m). Now assume (p_{j+1}). The same reasoning as in the proof of lemma (1.1) in [24] shows, that there is a positive number $\varepsilon \in \mathbb{R}$, such that

$$\varphi_j := \frac{1}{1+\varepsilon}(\varepsilon\varphi_{j+1} + \psi_{X_j})$$

has property (p_j).

This finally proves the theorem.

<u>REMARK 17:</u> If we could prove theorem 14 for proper mappings $f : X \to Y$ of paracompact complex spaces, it would follow that X is (k+1)-complete if Y is 1-complete.

Using this theorem, we can prove the following convexity theorem, which generalizes the main result in [24].

<u>THEOREM 18:</u> Let F be a globally generated k-ample coherent sheaf on a compact complex space X. Then the linear fibre space $\pi : \mathrm{Lin}(F) \to X$ is (k+1)-convex.

PROOF: Since F is globally genrated, $L(\mathrm{Lin}(F))^*$ is globally generated on $\mathbb{P}(\mathrm{Lin}(F))$. Let $\tilde{X} := \mathbb{P}(\mathrm{Lin}(F))$, $L := L(\mathrm{Lin}(F))$ and let $V \subset H^O(\tilde{X}, L^*)$ generate F. Then we have a holomorphic map

$$f : \tilde{X} \to \mathbb{P}(V^*)$$

and a Cartesian diagram

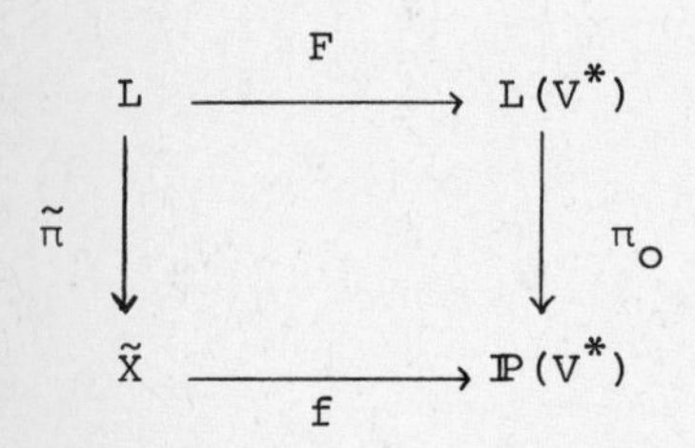

Using the canonical biholomorphic map

$$\alpha : \mathrm{Lin}(F) \setminus X \to L \setminus \tilde{X}$$

we see, that it suffices to show, that $\tilde{\pi} : L \to \tilde{X}$ is (k+1)-convex.

Now choose coordinates $z_o, \ldots, z_m$ in V^* and define a scalar product (,) on V^* by

$$(z, z') = \sum_{j=o}^{m} z_j \overline{z_j'} \; .$$

Let $\sigma : L(V^*) \to V^*$ denote the σ-process in $o \in V^*$. We get an induced scalar product $\sigma^*(\, , \,)$ on $L(V^*)$.

Now consider the natural local frame

$$s_i : U_i \to L(V^*) \times_{\mathbb{P}(V^*)} U_i$$

on the basis set $U_i = \{\langle z \rangle \in \mathbb{P}(V^*) \mid z_i \neq 0\}$ given by $s_i(\langle z \rangle) = (\langle z \rangle, \frac{1}{z_i} z)$.
The restriction g_i to U_i of the Fubini-metric g on $\mathbb{P}(V^*)$ is

$$g_i = \delta\bar{\delta} \log||\sigma \circ s_i||^2 .$$

Now pull back the scalar product on $L(V^*)$ to a scalar product H on L and let f^*s_i denote the induced frame on $f^{-1}U_i$. Then we compute $(h(l) := H(l,l) \; \forall \; l \in L)$

$$\delta\bar{\delta} \log h \circ f^*s_i = \delta\bar{\delta} \log||\sigma \circ s_i \circ f||^2 = f^*g_i$$

on $f^{-1}U_i$.

By assumption, f has fibres of dimension $\leq k$, so from theorem 14 we get a differentiable function φ on $\tilde{X}$, such that

$$L_\varphi + f^*g$$

has at most k non-positive eigenvalues for each point $\tilde{x} \in \tilde{X}$.

Define a new Finsler-metric h_φ on $\tilde{\pi} : L \rightarrow \tilde{X}$ by

$$h_\varphi(l) := e^{\varphi \, \tilde{\pi}} \cdot h .$$

Then on $f^{-1}U_i$

$$\delta\bar{\delta} \log(h_\varphi \circ f^*s_i) = (L_\varphi + f^*g)|_{f^{-1}U_i} .$$

Using a trivialization of L in $\tilde{x} \in f^{-1}U$ which is normal in $\tilde{x}$ with respect to the scalar product H_φ that induces h_φ we see, that h_φ is (k+1)-convex in $l \in L(\tilde{x})$ if and only if $\delta\bar{\delta} \log(h_\varphi \circ f^*s_i)$ has at most k non-positive eigenvalues in $\tilde{x}$. This proves the theorem.

<u>THEOREM 19:</u> Let $\pi : E \rightarrow X$ be a holomorphic vector bundle on a compact complex space X. If E is globally generated and k-ample, $\pi : E \rightarrow X$ is Finsler-(k+1)-positive.

<u>PROOF:</u> This follows from theorem 8 and theorem 18.

COROLLARY 20: Let $\pi : E \to X$ be a holomorphic vector bundle of rank r on a compact complex space X. If E is globally generated and k-ample, L(E) is (r+k)-positive.

PROOF: Use proposition 4.

3. COMPLEMENTS IN COMPLEX SPACES

Let X be a complex space, $A \subsetneq X$ a closed complex subspace, defined by the ideal $I \subset \mathcal{O}_X$. The generalized normal bundle of A in X is the linear fibre space $\pi : N_{A/X} \to A$ with [5]

$$N_{A/X} := \mathrm{Lin}(I) \times_X A = \mathrm{Lin}(I/_{I^2}).$$

If A is locally a complete intersection in X, the conormal sheaf $I/_{I^2}$ is locally free and we have [14]

$$I^K/_{I^{k+1}} = S^k(I/_{I^2}) .$$

We will show, that the complement $X\backslash A$ has a certain convexity property, if the normal bundle is sufficiently positive. Let us make this precise. Let $\varphi : X\backslash A \to \mathbb{R}$ be a differentiable function. For $x \in X\backslash A$ we denote the holomorphic tangent space to the hypersurface $\varphi = 0$ in x by

$$H(\varphi,x) = \{v \in T_X(x) \mid \delta\varphi(x)(v) = 0\}.$$

A differentiable function $\varphi : X\backslash A \to \mathbb{R}$ is an asymptotically q-convex exhaustion of $X\backslash A$ if it has the following properties:

i) $\lim_{n \to \infty} f(x_n) = \infty$ for any sequence $x_n \in X\backslash A$ with $\lim_{n \to \infty} x_n = a \in A$

ii) the restriction of the Levi-form $L_\varphi(x)$ to $H(\varphi,x)$ has at most $q-1$ non-positive eigenvalues for each $x \in X\setminus A$.

<u>THEOREM 21:</u> Let $A \subset X$ be a closed complex subspace with Finsler-$(k+1)$-positive normal bundle $N = N_{A|X}$. Then there exists an open neighbourhood U of A in X and an asymptotically (R_N+k)-convex exhaustion φ on $U\setminus A$. If in addition X is compact, the complement $X\setminus A$ is (R_N+k)-convex.

<u>PROOF:</u> Let $I \subset \mathcal{O}_X$ be the ideal defining $A \subset X$. We blow up X along A and get the following situation:

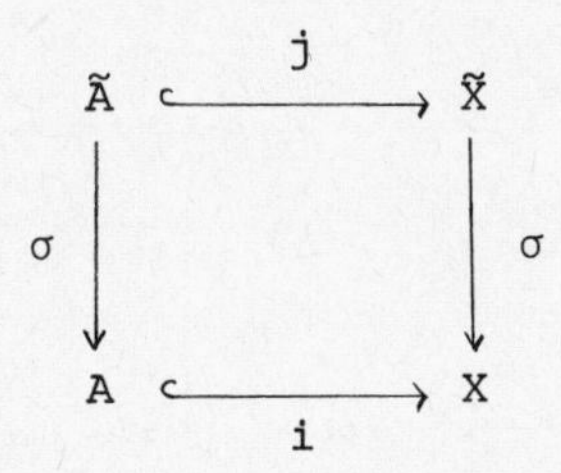

Here $\tilde{A} \subset \tilde{X}$ is a hypersurface with

$$\tilde{A} = \mathbb{P}(N) \cap \tilde{X} \subset \mathbb{P}(\mathrm{Lin}\ I).$$

The normal bundle of $\tilde{A}$ in $\tilde{X}$ is

$$N_{\tilde{A}/\tilde{X}} = L(N) \times_{\mathbb{P}(N)} \tilde{A}.$$

From proposition 3 and proposition 4 we see, that $N_{\tilde{A}/\tilde{X}}$ is (R_N+k)-positive. The claim now follows from the proof of the main result in [6] using the isomorphism

$$\tilde{X}\setminus\tilde{A} \underset{\sigma}{\to} X\setminus A .$$

<u>COROLLARY 22:</u> Assume that $A \subset X$ is compact, $N = N_{A/X}$ a globally generated vector bundle, such that its dual N^* is $(k+1)$-convex. Then there exists a relatively compact open neighbourhood U of A in X and a proper

differentiable function $\varphi : U\setminus A \to \mathbb{R}$ which is (R_N+k)-convex and satisfies $\lim_{n\to\infty} \varphi(x_n) = \infty$ for $x_n \to a \in A$. If in addition X is compact, the complement $X\setminus A$ is (R_N+k)-convex.

PROOF: From theorem 8 we see, that N is Finsler-(k+1)-positive. Using theorem 21 above we find an asymptotically (R_N+k)-convex exhaustion ψ of $X\setminus A$. Now the result follows from lemma (0.3.3) in [22].

COROLLARY 23: Assume that $A \subset X$ is compact, $N = N_{A/X}$ a globally generated, k-ample vector-bundle. Then there exists a relatively compact open neighbourhood U of A in X and a proper (R_N+k)-convex function $\varphi : U\setminus A \to \mathbb{R}$ with $\lim_{n\to\infty} \varphi(x_n) = \infty$ for $x_n \to a \in A$. If in addition X is compact, the complement $X\setminus A$ is (R_N+k)-convex.

PROOF: This follows from theorem 19 and theorem 21.

REMARK 24: i) Since r-positive vector bundles are in particular Finsler-r-positive, theorem 21 generalizes Satz 2 in [18].

ii) In the case $k = 0$, X compact, theorem 21 reduces to theorem 2.2 in [6].

iii) the corollaries above generalize proposition (0.4.6) and corollary (1.4) in [24].

iv) If $A \subset X$ is a closed submanifold of a rational homogeneous projective manifold X the ampleness of $N_{A/X}$ is bounded by the ampleness of the tangent bundle T_X. The ampleness of T_X for such homogeneous manifolds X has been determined by Goldstein [8].

We mention a few application of the previous results:

THEOREM 25: Let $A \subset X$ be a closed complex subspace in a compact complex space X. If the normal bundle $N = N_{A/X}$ of A in X is Finsler-(k+1)-positive, then for every coherent sheaf F on $X\backslash A$ we have

$$\dim_{\mathbb{C}} H^i(X\backslash A, F) < \infty \quad \text{for} \quad i \geq R_N+k.$$

<u>PROOF:</u> Theorem 21 and theorem 14 in [1].

<u>PROPOSITION 26:</u> Let $A \subset X$ be a compact complex subspace in a complex space X, F a coherent analytic sheaf on X. If the normal bundle $N = N_{A/X}$ is Finsler-(k+1)-positive, there exists a fundamental system $U_n(A)$ of neighbourhoods of A in X with the following properties:

i) $\dim_{\mathbb{C}} H^i(U_n(A), F) < \infty$ for $i < \operatorname{dih} F - (R_N+k)$

ii) the restriction map $H^i(U_n(A), F) \to H^i(A, F|_A)$ is bijective for $i < \operatorname{dih} F - (R_N+k)$ and injective for $i = \operatorname{dih} F - (R_N+k)$.

<u>PROOF:</u> This follows from theorem 21 and proposition (0.6) in [22].

For another application consider the following situation (S):

X is a rational <u>homogeneous</u> projective manifold, $A, B \subset X$ closed complex subspaces such that the normal bundle $N = N_{B/X}$ of B in X is Finsler-(k+1)-positive. Assume, that every component of A and B has dimension $\geq q := R_N+k$.

<u>THEOREM 27:</u> In the situation (S) the following holds:

i) B is connected.

ii) $H^i(X, B; \mathbb{C}) = 0$ for $i \leq \operatorname{dih} \mathcal{O}_B - q+1$.

iii) $H^i(A, A \cap B; \mathbb{C}) = 0$ for $i \leq \min(\dim A, \operatorname{dih} \mathcal{O}_B+1) - q$ if A is locally a complete intersection and $A\backslash A \cap B$ smooth.

THEOREM 28: In the situation (S) let F be a coherent sheaf on A. Then the following holds:

i) $\dim_{\mathbb{C}} H^i(B \cap A, F|_{B \cap A}) < \infty$ for $j < \text{dih } F - q$.

ii) the restriction $H^i(A,F) \to H^i(B \cap A, F|_{B \cap A})$ is bijective for $i < \min(\text{dih } F \ \text{dih } \mathcal{O}_B+1) - q$ and injective for $i = \min(\text{dih } F, \text{dih } \mathcal{O}_B+1) - q$.

iii) $H^i(A \setminus A \cap B, F) = 0$ for $i \geq \max(0, \dim A - \text{dih } \mathcal{O}_B - 1) + q$ if A is locally a complete intersection.

PROOF: From theorem 21 we get an asymptotically q-convex exhaustion in $U \setminus B$, where $U \subset X$ is an open neighbourhood containing B. The results then follow from § 3 in [22].

We will deal with the homotopy analog of the Lefschetz-type theorem 27 in another paper.

REFERENCES

[1] Adreotti, A. and H. Grauert: Théorèmes de finitude pour la cohomologie des espaces complexes. Bull. Soc. Math. France 90, 193 - 259 (1962).

[2] Barth, W.: Der Abstand von einer algebraischen Mannigfaltigkeit im komplex-projektiven Raum. Math. Ann. 187, 150 - 162 (1970).

[3] Barth, W.: Transplanting cohomology classes in complex-projective space. Amer. Journ. of Math. 92, 951 - 967 (1970).

[4] Fischer, G.: Complex analytic geometry. LNM 538. Springer Verlag, Berlin-Heidelberg-New York (1976).

[5] Fritzsche, K.: q-konvexe Restmengen in kompakten komplexen Mannigfaltigkeiten. Math. Ann. 221, 251 - 273 (1976).

[6] Fritzsche, K.: Pseudoconvexity properties of complements of analytic subvarieties. Math. Ann. 230, 107 - 122 (1977).

[7] Fulton, W. and R. Lazarsfeld: Connectivity and its applications in algebraic geometry. LNM 862, Springer-Verlag, Berlin-Heidelberg-New York (1981).

[8] Goldstein, N.: Ampleness and connectedness in complex $G/_P$. Trans. Amer. Math. Soc. 274, 361 - 373 (1982).

[9] Grauert, H.: Über Modifikationen und exzeptionelle analytische Mengen. Math. Ann. 146, 331 - 368 (1962).

[10] Grauert, H. and O. Riemenschneider: Verschwindungssätze für analytische Kohomologiegruppen auf komplexen Räumen. Invent. Math. 11, 263 - 292 (1970).

[11] Griffiths, P.: Hermitean differential geometry, Chern classes and positive vector bundles. In: Global Analysis. Papers in honor of K. Kodaira. (New York. Princeton Univ. Press (1969).

[12] Hartshorne, R.: Ample vector bundles. Inst. Hautes Études Sci. Publ. Math. 29, 63 - 94 (1966).

[13] Hartshorne, R.: Cohomology of non-complete algebraic varieties. Compos. Math. 23, 257 - 264 (1971).

[14] Hartshorne, R.: Ample subvarieties of algebraic varieties. LNM 156, Springer Verlag, Berlin-Heidelberg-New York (1970).

[15] Hironaka, H.: Bimeromorphic smoothing of complex analytic space. Math. Inst. Warwick Univ. England (1971).

[16] Kobayashi, S.: Negative vector bundles and complex Finsler structures. Nagoya Math. Journ. 57, 153 - 166 (1975).

[17] Riemenschneider, O.: Characterizing Moisezon spaces by almost positive coherent sheaves. Math. Z. 123, 263 - 284 (1971).

[18] Schneider, M.: Über eine Vermutung von Hartshorne. Math. Ann. 201, 221 - 229 (1973).

[19] Schneider, M.: Tubenumgebungen Steinscher Räume. Manuscripta math. 18, 391 - 397 (1976).

[20] Sommese, A.: Submanifolds of Abelian varieties. Math. Ann. 233, 229 - 256 (1978).

[21] Sommese, A.: Concavity theorems. Math. Ann. 235, 37 - 53 (1978).

[22] Sommese, A.: Complex subspaces of homogeneous complex manifolds. I. Transplanting theorems. Duke Journ. of Math. 46, 527 - 548 (1979).

[23] Sommese, A.: Complex subspaces of homogeneous complex manifolds. II. Homotopy results. Nagoya Math. Journ. 86, 101 - 129 (1982).

[24] Sommese, A.: A convexity theorem. Proc. of Symp. in Pure Math. on Singularities, Arcata 1981, Vol. 40, part 2, 497 - 505 (1983).

[25] Sommese, A. and A. Van de Ven: Homotopy groups of pullbacks of varieties. Preprint (1984).

[26] Spallek, K.: Differenzierbare Räume. Math. Ann. 180, 269 - 296 (1969).

SUBVARIETIES IN HOMOGENEOUS MANIFOLDS

Christian Okonek

This note is intended to supplement recent papers of Diederich/Fornaess [2], [3] and M. Peternell [12]. In these papers the authors study a class of functions which they call q-convex functions with corners.

A continuous function φ on a complex space is q-convex with corners if φ is locally the supremum of finitely many q-convex C^{∞}-functions. Those functions naturally occur if one studies complements of analytic subspaces in complex manifolds. This was already noticed by Grauert [7]. In [2] and [3] Diederich/Fornaess prove a smoothing result for such functions. If X is a complex space of dimension n, $\varphi : X \to \mathbb{R}$ a q-convex function with corners, then in an arbitrary neighborhood of φ there exists a differentiable function $\tilde{\varphi}$ which is $\tilde{q}$-convex, where $\tilde{q} = n - [\frac{n}{q}] + 1$. Here $[\frac{n}{q}]$ is the largest integer $\leq \frac{n}{q}$.

In order to apply this theorem, one needs a method to construct q-convex functions with corners. In [12] M. Peternell defines a condition (E_k') for every complex space X. A complex space X satisfies (E_k') if there exists an open neighborhood U of the diagonal $\Delta_X \subset X \times X$ and a continuous function

$$\varphi : U \setminus \Delta_X \to \mathbb{R}$$

which is $(n(X)+k)$-convex with corners, $n(X) := \inf_{x \in X}\{\dim_x X\}$, such that for every sequence $(x_m, x) \in U \setminus \Delta_X$ converging to (x,x) $\lim_{m \to \infty} \varphi(x_m,x) = \infty$. If a compact complex space X satisfies (E_k') and $A \subset X$ is a closed subspace of maximal codimension q, then the complement $X \setminus A$ carries a continuous function, which is $(q+k)$-convex with corners outside a compact set. If X is a compact homogeneous manifolds, there exists a $(q+k)$-convex function with corners on $X \setminus A$.

Unfortunately the condition (E_k') is sometimes hard to check [12].

In this paper we give a criterion (theorem 1) which implies (E_k') . Then we use the smoothing result of Diederich/Fornaess [3] and the stratified Morse theory of Goresky/MacPherson [6] to prove a Lefschetz-type theorem (theorem 3).

Recall from [8], that a vector bundle E on a complex space X is Finsler-$(k+1)$-positive, if there exists a Finsler-metric

$$\rho : E \setminus X \to (0,\infty) \qquad \rho(z \cdot e) = |z|^2 \rho(e)$$

which is strictly plurisubharmonic on fibres, such that the Levi-form $L_\rho(e)$ has at least $\operatorname{emb} \dim_x X - k$ negative eigenvalues for every $e \in E(x) \setminus \{0\}$, $x \in X$.

<u>Theorem 1:</u> A compact complex manifold X with a Finsler-$(k+1)$-positive tangent bundle satisfies (E_k') .

Proof: In [8] we proved a more general result: Let $A \subset X$ be a compact subspace in a complex space X which has a Finsler-$(k+1)$-positive normal bundle $N_{A/X}$ of rank q . Then there exists a relatively compact open neighborhood U of A in X and a proper $(q+k)$-convex C^∞-function

$$\varphi : U \setminus A \to (0,\infty)$$

with $\lim_{m\to\infty} \varphi(x_m) = \infty$ for every sequence $x_m \in U\setminus A$ converging to a point of A . Applying this result to the diagonal $\Delta_X \subset X \times X$ gives our criterion.

Clearly every (k+1)-positive vector bundle is Finsler-(k+1)-positive [8]. But there is a more useful condition:

A vector bundle E on a complex space is k-ample in the sense of Sommese, if some positive power of the universal quotient line bundle $\mathcal{O}_{\mathbb{P}(E)}(1)$ on $\mathbb{P}(E)$ is globally generated such that the associated morphism has only fibres of dimension $\leq k$ [13].

In [8] we proved, that on a compact space every globally generated k-ample vector bundle is Finsler-(k+1)-positive. So we have

Corollary 2: A compact complex manifold X satisfies (E_k') if the tangent bundle is globally generated and k-ample.

This corollary already follows from the results of Sommese [14]. The point is: k-ampleness is an algebro-geometric notion which can be calculated in many cases [13] .

The ampleness of the tangent bundle of a rational homogeneous projective manifold has been determined by Goldstein using root systems [5]. E. g. the tangent bundle of $\mathbb{P}^n$ is 0-ample.

Consider now the following situation (S):

$$(S)\quad \begin{array}{ccc} Y & \hookrightarrow & X \\ \cup & \square & \cup \\ Y \cap A & \hookrightarrow & A \end{array}$$

X is a compact homogeneous manifold with k-ample tangent bundle, A is a closed subspace of maximal codimension q, $Y \hookrightarrow X$ a closed subspace of pure dimension m.

Theorem 2: In the situation (S) the complement $Y \setminus Y \cap A$ is $(m - [\frac{m}{q+k}] + 1)$-complete.

Proof: Since the tangent bundle of a homogeneous manifold is globally generated, corollary 2 shows, that condition (E'_k) is true for X. From Peternell's paper [12] we get the existence of a (q+k)-convex function with corners φ on $X \setminus A$. Now apply the smoothing result of Diederich/Fornaess [3] to the restriction $\varphi|_{Y \setminus Y \cap A}$.

The 'théorème de finitude' of Andreotti/Grauert [1] gives vanishing theorems as usual.

Let $\tilde{\varphi} : Y \setminus Y \cap A \to \mathbb{R}$ be a $(m - [\frac{m}{q+k}] + 1)$-convex exhaustion function. We approximate $-\tilde{\varphi}$ by a Morse function ψ relative to a stratification (S_α) of $(Y, Y \cap A)$. It is easy to see, that the index $i_y(\psi)$ of the restriction $\psi|_{S_\alpha}$ of ψ to the stratum S_α which contains the critical point y is bounded below by

$$i_y(\psi) \geq [\frac{m}{q+k}] .$$

Let Γ_ψ denote the convexity deficiency of ψ [6] . As in [8] one calculates

$$\Gamma_\psi \leq m - [\frac{m}{q+k}] .$$

For any stratum S_α the local defect at S_α is the number

$$\Delta_\alpha := \dim L_\alpha - \inf\{i | \pi_i(L_\alpha) \neq 0\} ,$$

where L_α is the complex link of S_α [6] .

Let

$$\Delta_{Y \setminus Y \cap A} := \sup_{S_\alpha \subset Y \setminus Y \cap A} \{\Delta_\alpha\} .$$

Using the stratified Morse theory of Goresky/MacPherson [6] we obtain

Theorem 3: In the situation (S)

$$\pi_j(Y, Y \cap A; *) = 0 \qquad j \leq [\frac{m}{q+k}] - 1 - \Delta_{Y \setminus Y \cap A} .$$

Corollary 4: If in the situation (S) $Y \setminus Y \cap A$ is a local complete intersection, then

$$\pi_j(Y, Y \cap A; *) = 0 \qquad j \leq [\frac{m}{q+k}] - 1 .$$

Compare [2], [3], [4], [10], [11] for related results.

References

[1] Andreotti, A. and H. Grauert: Théorèmes de finitude pour la cohomologie des espaces complexes. Bull. Soc. Math. France 90, 193-259 (1962).

[2] Diederich, K. and I. E. Fornaess: Smoothing q-convex functions and vanishing theorems. To appear Invent. math. (1985).

[3] Diederich, K. and I. E. Fornaess: Smoothing q-convex functions in the singular case. Preprint (1985).

[4] Faltings, G.: Über lokale Kohomologiegruppen hoher Ordnung. J. reine und angew. Math. 313, 43-51 (1980)

[5] Goldstein, N.: Ampleness and connectedness in complex G/P. Trans. Amer. Math. Soc. 274, 361-373 (1982)

[6] Goresky, M. and R. MacPherson: Stratified Morse theory. Forthcoming book.

[7] Grauert, H.: Kantenkohomologie. Comp. Math. 44, 79-101 (1981)

[8] Okonek, C.: Concavity, convexity and complements in complex spaces. Preprint (1985)

[9] Okonek, C.: Barth-Lefschetz theorems for singular spaces. Preprint (1985)

[10] Peternell, M.: Ein Lefschetz-Satz für Schnitte in projektiv algebraischen Mannigfaltigkeiten. Math. Ann. 264, 361-388 (1983)

[11] Peternell, M.: Homotopie in homogenen komplexen Mannigfaltigkeiten. Math. Z. 188, 271-278 (1985)

[12] Peternell, M.: Continuous q-convex exhaustion functions. Preprint (1985)

[13] Sommese, A.J.: Submanifolds of Abelian varieties. Math. Ann. 233, 229-256 (1978)

[14] Sommese, A.J.: A convexity theorem. Proc. of Symposia in Pure Mathematics (Arcata 1981) Vol. 40 Part 2 497-505(1983)

Sonderforschungsbereich 170: Geometrie und Analysis

Universität Göttingen

Bunsenstr. 3 - 5

D-3400 Göttingen F.R.G.

RATIONAL CURVES IN MOIŠEZON 3-FOLDS

Thomas Peternell
Mathematisches Institut der
Universität Bayreuth
Postfach 3008
8580 Bayreuth
Bundesrepublik Deutschland

Introduction

Since the fundamental work of S. Mori on manifolds whose canonical bundles are not numerically effective ([2]) one knows that rational curves have a great influence on the geometry of projective manifolds. If X is such a manifold - say of dimension 3 - and C a rational curve such that $(c_1(K_X)\cdot C) < 0$ (K_X the canonical bundle) then one can choose C "extremal" in some sense and C gives either rise to a modification collapsing exactly C and all homologous curves or to a fibering over a lower - dimensional manifold with the same "collapsing behaviour".
In this paper we investigate Moišezon 3-folds X which are not projective. While in a "general" projective manifold there will be no rational curve, we show that any non-projective X carries some rational curve (sect. 2). This curve may come either from a modification or fibering as above or may simply be created by the non-projectiveness of X.
The method of proof is to take a "projectivization" $\hat{X}$ of X, i.e. we blow up smooth subvarieties of X several times. Then we investigate the (extremal) rational curves on $\hat{X}$ and show that not every rational curve on $\hat{X}$ can be contracted by $\pi : \hat{X} \longrightarrow X$. So the images of those curves are rational curves in X. The character of the constructed rational curves C in X may be very different. C may be rigid or may move in a family or be contained in a rational surface. If K_X is "very positive", there will only a few rigid curves C, may be only one coming from an extremal rational curve on $\hat{X}$ which is not numerically effective and thus defining a modification $\phi : \hat{X} \longrightarrow Y$. In this case the exceptional divisor $D \subset \hat{X}$ of ϕ satisfies $\dim \pi(D) = 1$.
In the proof we will need a special projectivization $\pi : \hat{X} \longrightarrow X$, called "standard". This projectivization is contructed in [5] and in sect. 1 we recall the facts we need. The philosophy is to recognize

certain bad curves in X which prevent the manifold to be projective. For instance, curveswhich are homologous to O, are bad. Roughly speaking, the bad curves are those to be blown up. One could expect that the bad curves are all rational - this would give a deeper understanding of the non-projectiveness of X. A first step in this direction is the following problem first posed by Moišezon :
Let $\hat{X}$ a projective 3-fold, X a non-projective Moišezon 3-fold, and assume that

$$\pi : \hat{X} \longrightarrow X \text{ is the blow-up in the smooth curve } S \subset X.$$

Is $S \simeq \mathbb{P}_1$?
In section 4 we give an affirmative answer in the case $(c_1(K_X)\cdot S) = O$ and we show that $(c_1(K_X)\cdot S) < O$ is not possible.
The case $(c_1(K_X)\cdot S) > O$ seems to be rather hard. For example, if the ruled surface $\hat{S} = \pi^{-1}(S)$ is unstable, then the normal bundle of S is negative and hence S exceptional.

§ 1 Bad curves in Moišezon 3-folds

In this section we review some of the result of [5], sect. 3.
X denotes always a Moišezon 3-fold. We want to find a rather explicit projective desingularization of X.

Notation

A projectivization of X is a modification $\pi : \hat{X} \longrightarrow X$ where π is a finite sequence of blow-up's with smooth centers such that $\hat{X}$ is projective. Also $\hat{X}$ is called projectivization of X.

Definition

1) An almost ample line bundle on X is a triple (G,F,π) consisting of
   a) a projectivization $\pi : \hat{X} \longrightarrow X$
   b) $F \in \mathrm{Pic}(\hat{X})$ ample
   c) $G \in \mathrm{Pic}(X)$ such that $G = \pi_*(F)^{**}$.
2) A projectivization π is called minimal if in an obvious sense no π_ν can be omitted and if the degeneracy set $S \subset X$ of π is minimal

with these properties (i.e. there is no projectivization π' of X such that no π'_μ can be omitted and such that $S' \subsetneq S$ (S' the degeneracy set of π').

Definition

An irreducible curve $C \subset X$ is called bad of type I with respect to a projectivization $\pi : \hat{X} \longrightarrow X$ (for short: π-bad of type I) iff:

a) there is a current $T = \lim_{j\to\infty} \sum_{i=1}^{n_j} \lambda_{ij} T_{C_{ij}}$,

   all $\lambda_{ij} \geq 0$, $T_{C_{ij}}$ denoting integration over the irreducible curve $C_{ij} \subset X$ and convergence takes place in the weak topology, such that $C + T \sim 0$ (homological equivalence).

b) for all almost ample (G,F,π) :

$$\left(c_1(G) \cdot C \right) \leq 0.$$

   C is called bad of type I iff there is a π such that C is bad of type I with respect to π.

Definition

An irreducible curve $C \subset X$ is called bad of type II with respect to a projectivization $\pi : \hat{X} \longrightarrow X$ iff:

a) 3.2.a holds for C
b) C is not π-bad of type I
c) there exists an almost ample line bundle (G,F,π) such that for all $\mu \in \mathbb{N}$ the vector space $\operatorname{Im}\{H^0(X,G^\mu) \longrightarrow H^0(C,G^\mu(C)\}$ does not define a bimeromorphic map

$$C \longrightarrow \mathbb{P}_N .$$

Theorem 1:

If X has no bad curves of type I, then X is projective.

Theorem 2:

There exists a finite sequence of blow-up's $\pi : \hat{X} \longrightarrow X$ such that
a) $\hat{X}$ is procetive
b) any center is either a smooth point or a smooth bad curve of type I.
More precisely, one can choose π as follows. Fix a minimal projectivization $\rho : \tilde{X} \longrightarrow X$. First make all ρ-bad curves disjoint and smooth (by blowing up points).
So we may assume a priori disjointness.
Take a bad curve of type I and blow it up. If the resulting ruled surface R is stable (i.e. the invariante $e \leq 0$ (see [1], chap 5, sect. 2) stop and take the next bad curve of type I.
If R is unstable ($e>0$), the negative section of R may be bad of type I. In this case blow up the negative section and so on. This procedure terminates.

Remark: A bad curve of type I is rigid - this being wrong in general for bad curves of type II.

§ 2 Rational curves on Moišezon 3-folds

Theorem: Let X be a Moišezon 3-fold. If X does not contain a rational curve, X is projective.

Proof: Assume that X is not projective.

Let $\pi : \hat{X} \xrightarrow{\pi_r} X_{r-1} \longrightarrow \ldots \longrightarrow X_1 \xrightarrow{\pi_1} X$ be a finite sequence of blow-up's such that
a) the center S_i of π_i is smooth ($1 \leq i \leq r$),
b) $\hat{X}$ is projective,
c) π is standard.
Let $S = \{x \in X |\ \dim \pi^{-1}(x) \geq 1\}$ the degeneracy set of π, $\hat{S} := \pi^{-1}(S)$.
Clearly the canonical bundle $K_{\hat{X}}$ is not numerically effective (nef) (e.g. if C is an irreducible curve in $\hat{X}$ with $\dim \pi_r(C) = 0$, then $K_{\hat{X}}|C$ is negative).
By Mori [2] there exists a holomorphic surjective map $\phi : \hat{X} \longrightarrow Y$ with certain properties which will be explained when used.

a) First we treat the case $\dim Y \leq 2$.
Then Y is smooth by Mori.
$\dim Y = 0$ is a priori not possible since by Mori we would have $\rho(\hat{X}) = 1$ (Picard number of $\hat{X}$) which is clearly impossible ($\rho(\hat{X}) > \rho(X) \geq 1$).
Assume that $\dim Y = 2$.
By [2], $K_{\hat{X}}^{-1}$ is ϕ-ample, hence the generic fiber $\phi^{-1}(y)$ is $\mathbb{P}_1$ and since a generic $\phi^{-1}(y)$ intersects $\hat{S}$ in at most finitely many points, $\pi(\phi^{-1}(y))$ is a rational curve in X, contradiction.
Now assume $\dim Y = 1$.
Let F be a generic smooth fiber of ϕ.
Since $K_{\hat{X}}^{-1}$ is again ϕ-ample, K_F must be negative. Since $\dim(F \cap \hat{S}) \leq 1$ (F generic), $F \cap \hat{S}$ consists of at most finitely many curves. The negativity of K_F implies that F is rational or birational to a ruled surface. In any case, F has ∞ many rational curves. Hence there is a rational curve $C \subset F$ such that $C \not\subset \hat{S}$.
Its image is a rational curve in X, contradiction.

b) Now let $\dim Y = 3$. Then ϕ is a modification, more precisely, there exists an irreducible hypersurface $D \subset \hat{X}$ such that $\phi : \hat{X} \setminus D \longrightarrow Y \setminus \phi(D)$ is biholomorphic and ϕ is the blow-up of $\phi(D)$. Moreover,

α) if $\dim \phi(D) = 1$, Y and $\phi(D)$ are smooth, so D is a ruled surface,

β) if $\dim \phi(D) = 0$, $D \simeq \mathbb{P}_2$ or $D \simeq \mathbb{P}_1 \times \mathbb{P}_1$ or D is an irreducible singular quadric in $\mathbb{P}_3$.

b_1) First assume $\dim \pi(D) = 2$.
Again $\dim (D \cap \hat{S}) \leq 1$ and on the other hand D has ∞ many rational curves, so we are done.

b_2) $\dim \pi(D) = 1$.
Let $\hat{S}_\nu$ be the strict transform of S_ν in $\hat{X}$. Then the $\hat{S}_\nu$, $1 \leq \nu \leq r$, are exactly the irreducible components of $\hat{S}$.
Because of $\dim \pi(D) = 1$ there is an uniquely determined ν such that $D = \hat{S}_\nu$.
Necessarily $\dim S_\nu = 1$. So S_ν is a smooth curve in $X_{\nu-1}$ and $\hat{S}_\nu$ is a ruled surface or a ruled surface sometimes blown up.
Hence $\hat{S}_\nu$ cannot be $\mathbb{P}_2$ or a quadric, so $D = \hat{S}_\nu$ is ruled (by Mori).
For $\mu \geq \nu$ let $S_{\nu\mu}$ be the strict transform of $\pi_\nu^{-1}(S_\nu)$ in X_μ. Then for $\mu > \nu$ either $S_{\nu,\mu-1} \cap S_\mu = \emptyset$ or $S_\mu \subset S_{\nu,\mu-1}$ and $\dim S_\mu = 1$ (otherwise $\hat{S}_\nu$ would not be ruled).
Assume first that there is some $\mu_o > \nu$ with $S_{\mu_o} \subset S_{\nu,\mu_o-1}$.
Assume furthermore $\dim \phi(D) = 1$.

Let F be a fiber of $\phi|D$.

Then (*) $(D\cdot F)_{\hat{X}} = -1$.

Let F' be a fiber of $\pi_\nu|S_{\nu,\nu} \longrightarrow S_\nu$.

Then $(S_{\nu,\nu}\cdot F')_{X_\nu} = -1$ (**).

If the strict transform $\hat{F}'$ of F' in $\hat{X}$ is not a fiber of $\phi|D$, then D has two rulings, hence $D \simeq \mathbb{P}_1 \times \mathbb{P}_1$. So $S_\nu = \mathbb{P}_1$ and $\pi_1 \circ \ldots \circ \pi_{\nu-1}(S_\nu)$ is a rational-curve in X.

So we may assume $\hat{F}' = F$.

The form of the standard projectivization implies that either there is no $\mu>\nu$ such that S_μ is a fiber of $S_{\nu,\mu-1} \longrightarrow S_\nu$ or if some S_{μ_1} is a fiber of S_{ν,μ_1-1}, then there is some μ_2 such that $S_{\mu_2} \subset S_{\nu,\mu_2-1}$ and S_{μ_2} is not a fiber of $S_{\nu,\mu_2-1} \longrightarrow S_\nu$.

Then an easy calculation of intersection numbers shows:

$$(D\cdot F)_{\hat{X}} < (S_{\nu,\nu} \cdot F')_{X_\nu}$$

contradicting (*) and (**).

If dim $\phi(D) = 0$, $D = \mathbb{P}_1 \times \mathbb{P}_1$ and $\pi(D)$ is a rational curve in X.

So we may assume that $S_\mu \cap S_{\nu,\mu-1} = \emptyset$ for all $\mu>\nu$, i.e. $\nu=r$ (w.l.o.g.). If dim $\phi(D) = 1$, necessarily $D = \mathbb{P}_1 \times \mathbb{P}_1$ since otherwise D could have only one ruling, hence $\phi = \pi_r$ and $Y = X_{r-1}$ contradicting the minimality of r (Y is projective).

If dim $\phi(D) = 0$, a priori $D = \mathbb{P}_1 \times \mathbb{P}_1$. In both cases we get a rational curve in X.

So we are reduced to the last case:

$b_3)$ dim $\pi(D) = 0$.

As in b_2) we have some ν such that $D = \hat{S}_\nu$. First assume that dim $S_\nu = 0$, so $S_{\nu,\nu} = \pi_\nu^{-1}(S_\nu) \simeq \mathbb{P}_2$. If we blow up a point in our standard projectivization, we do this either in order to smooth a curve or in order to separate at least two curves (which must be blown up later). In both cases it follows that $\hat{S}_\nu$ is $\mathbb{P}_2$ blown up in at least two points, so $\hat{S}_\nu$ cannot be ruled, $\mathbb{P}_2$ or a quadric contradicting $D = \hat{S}_\nu$.

So we must habe dim $S_\nu = 1$ but $\dim(\pi_1 \cdot \ldots \cdot \pi_{\nu-1})(S_\nu) = 0$.

As in case b_2), $D = \hat{S}_\nu$ must be ruled. Furthermore we may either assume $\nu = r$ or there is some $\mu_o>\nu$ such that $S_{\mu_o} \subset S_{\nu,\mu_o-1}$.

$b_{3.1})$ Assume first the existence of μ_o.

Let dim $\phi(D) = 1$. Define F and F' as in b_2). If the strict transform $\hat{F}'$ of F' in $\hat{X}$ is not a fiber F, then $D = \mathbb{P}_1 \times \mathbb{P}_1$.

Remark that $D \simeq S_{\nu,\mu}$ for all $\mu>\nu$.
Then no bad curve of type I can be contained in any $S_{\nu,\mu}(\mu>\nu)$, compare [4], p. 421. Hence μ_o cannot exist. So $F = \hat{F}'$. Then we can argue exactly as in the analogous case in b_2) and get a contradiction. Now assume $\dim \phi(D) = 0$. Then $D = \mathbb{P}_1 \times \mathbb{P}_1$. Then even $s\times\mathbb{P}_1 \sim \mathbb{P}_1 \times t$ for all $s,t \in \mathbb{P}_1$ (on $\hat{X}$, hence on X_μ), so (as above) no bad curve of type I can exist in $S_{\nu,\mu}$ $(\mu>\nu)$, contradiction.

$b_{3.2}$) We come down to the case $\nu=r$. So $D = \mathbb{P}_1 \times \mathbb{P}_1$.
There is some $\mu<\nu$ such that

$$\dim \left(\pi_{\mu+1} \cdot \ldots \cdot \pi_r \right) \left(\hat{S}_r \right) = 1$$

$$\text{but } \dim \left(\pi_{\mu} \cdot \ldots \cdot \pi_r \right) \left(\hat{S}_r \right) = 0.$$

So either $S_{r\mu} := (\pi_{\mu+1} \cdot \ldots \cdot \pi_r)(\hat{S}_r)$ is a fiber F_μ of π_μ or $S_{r\mu} \subset \pi_\mu^{-1}(S_\mu) = \mathbb{P}_2$ ($\dim S_\mu = 0$).
Note that $S_r \sim 0$ is impossible in both cases, so necessarily $\dim \phi(D) = 1$!
This implies easily for the normal bundle of S_r:

$$N_{S_r|X_{r-1}} = \mathcal{O}(-1) \oplus \mathcal{O}(-1).$$

First assume that $S_{r\mu} = F_\mu$.

Hence $N_{S_{r\mu}|X_\mu} = \mathcal{O} \oplus \mathcal{O}(-1)$.

Now we compute the normal bundle of $S_{r,\mu+1}$ in $X_{\mu+1}$. This is only interesting if $S_{\mu+1} \cap F_\mu \neq \emptyset$ (and clearly we may assume that this holds). By our standard projectivization, $S_{\mu+1} \neq F_\mu$.
If $\dim S_{\mu+1} = 0$,

$$N_{S_{r,\mu+1}|X_{\mu+1}} = \mathcal{O}(-1) \oplus \mathcal{O}(-2)$$

(consider the exact sequence of normal bundles

$$0 \longrightarrow N_{S_{r,\mu+1}|S_{\mu,\mu+1}} \longrightarrow N_{S_{r,\mu+1}|X_{\mu+1}} \longrightarrow N_{S_{\mu,\mu+1}|X_{\mu+1}}|S_{r,\mu+1} \longrightarrow 0 \;)$$

If $\dim S_{\mu+1} = 1$ and $S_{\mu+1} \not\subset \pi_\mu^{-1}(S_\mu)$, we compute in this manner:

$$N_{S_{r,\mu+1}|X_{\mu+1}} = \mathcal{O}(-1) \oplus \mathcal{O}(-2) .$$

If $S_{\mu+1} \subset \pi_\mu^{-1}(S_\mu)$ (then by our standard projectivization $e(S_{\mu,\mu})>0$ and $S_{\mu+1} \subset S_{\mu,\mu}$ is the negative section),

$$N_{S_{r,\mu+1}|X_{\mu+1}} = \mathcal{O} \oplus \mathcal{O}(-2) \text{ or } \mathcal{O}(-1) \oplus \mathcal{O}(-1).$$

Now we continue computing the normal bundles

$$N_{S_{r,\kappa}|X_\kappa}, \ \kappa > \mu+1 .$$

Except for the case $S_{\mu+1} \subset \pi_\mu^{-1}(S_\mu)$ (dim $S_{\mu+1} = 1$), we get: $c_1(N_{S_r|X_{r-1}}) \leqq -3$ yielding a contradiction:
But if $S_{\mu+1} \subset \pi_\mu^{-1}(S_\mu)$ we have

$$c_1(N_{S_{r,\mu+1}|X_{\mu+1}}) = -2 \quad ,$$

and sometimes it must happen:

$$S_{r,\kappa} \not\subset S_{\mu,\kappa}$$

(otherwise S_r could not be bad of type I), so also here we finally get

$$c_1(N_{S_r|X_{r-1}}) \leqq -3 \quad .$$

So we have dim $S_\mu = 0$, so $S_{r_\mu} \subset \pi_\mu^{-1}(S_\mu) = \mathbb{P}_2$. Let d be the degree of $S_{r_\mu} \subset \mathbb{P}_2$. Let $\tilde{\mathbb{P}}_2$ be the strict transform of $\mathbb{P}_2 = \pi_\mu^{-1}(S_\mu)$ in X_{r-1}. We have the exact sequence

$$\text{(S)} \qquad 0 \longrightarrow N_{S_r|\tilde{\mathbb{P}}_2} \longrightarrow N_{S_r|X_{r-1}} \longrightarrow N_{\tilde{\mathbb{P}}_2|X_{r-1}}|S_r \longrightarrow 0$$

$\alpha := c_1(N_{\tilde{\mathbb{P}}_2|X_{r-1}}|S_r) = (\tilde{\mathbb{P}}_2 \cdot S_r)_{X_{r-1}} <$
$(\mathbb{P}_2 \cdot S_{r\mu})_{X_\mu} = -d$ (because $(\mathbb{P}_2 \cdot S_{r\mu})_{X_\mu}$

$$= c_1(N_{\mathbb{P}_2}|X_\mu \, |S_{r\mu})$$

$$= c_1(\mathcal{O}_{\mathbb{P}_2}(-1)|S_{r\mu}))$$

$$c_1(N_{S_r}|\widetilde{\mathbb{P}}_2) = \mathcal{O}(d') \text{ for some } d'.$$

Now (S) is just the sequence

$0 \longrightarrow \mathcal{O}(d') \longrightarrow \mathcal{O}(-1)\oplus\mathcal{O}(-1) \longrightarrow \mathcal{O}(\alpha) \longrightarrow 0.$

This implies $d' \leq -1$, hence $\alpha \geq -1$ contradicting $\alpha < -d \leq -1$.

This completes the proof of the theorem.

§ 3 On the structure of bad curves

In this section we assume always the following situation. X is a non-projective Moišezon 3-fold, $S \subset X$ a smooth curve, $\pi : \hat{X} \longrightarrow X$ the blow-up of S and $\hat{X}$ is projective.

We want to investigate the structure of S.

$\hat{S} := \pi^{-1}(S)$ is a ruled surface. We can write $\hat{S} = \mathbb{P}(\mathcal{E})$ where $\mathcal{E}$ is a so-called normalized rank-two-bundle on S, i.e.:

$$H^0(S,\mathcal{E}) \neq 0, \text{ but } H^0(S,\mathcal{E}\otimes\mathcal{L}) = 0$$

for any $\mathcal{L} \in \mathrm{Pic}(S)$ with $\deg(\mathcal{L}) < 0$. Let $e := -c_1(\mathcal{E})$. e is an important invariant of $\hat{S}$ (see [1]). We always fix a section $\widetilde{S} \subset \hat{S}$ such that $\mathcal{O}([\widetilde{S}]) \simeq \mathcal{O}_{\mathbb{P}(\mathcal{E})}(1)$. Then $\widetilde{S}^2 = -e$. All what we need on ruled surfaces can be found in Hartshorne [1].

Lemma 1: $\quad (\widetilde{S}\cdot\hat{S})_{\hat{X}} < 0.$

Proof: (compare [4], p. 409) : It is sufficient to the proof the existence of $\mathcal{L} \in \mathrm{Pic}(\hat{X})$ ample such that $(c_1(\pi_*(\mathcal{L}))\cdot S) \leq 0$.

But let $\mathcal{L}$ be any ample line bundle in $\hat{X}$. By [6], there exists a positive current

$$T = \lim_{j\to\infty} \sum_{i=1}^{n_j} \lambda_{ij} \, T_{C_{ij}} \quad \text{with } S+T\sim 0,$$

where $\lambda_{ij} > 0$, $T_{C_{ij}}$ is the current obtained by integration over the irreducible curve $C_{ij} \neq S$ and convergence is taken in the weak topology.

Consequently:

$$\left(c_1(\pi_*(\mathcal{L})) \cdot [T] \right) \geq 0 \quad \text{and hence}$$

$$\left(c_1(\pi_*(\mathcal{L})) \cdot S \right) \leq 0.$$

Lemma 2:

Let $\hat{N}^*$ be the conormal bundle of $\hat{S}$ in $\hat{X}$ and write $\hat{N}^* \sim \tilde{S} + bF$ for numerical equivalence where F is a fiber of $\pi|\hat{S} \longrightarrow S$. Then

1) $a = 1$
2) $b > e$
3) $c_1(N) = e - b$
4) $2 - 2g + c_1(K_X|S) = 2b - e$

g being the genus of S.

Proof: 1) is clear, 2) is just lemma 1. 3) $(\hat{N}^*)^2_{\hat{S}} = 2b - e$ (see [1]), on the other hand $(\hat{N}^*)^2_{\hat{S}} = c_1(N^*)$. 4) This is the adjunction formula plus 3).

Theorem 3:

If $c_1(K_X|S) = 0$, then $S \simeq \mathbb{P}_1$ and $N_{S|X} \simeq \mathcal{O}(-1) \oplus \mathcal{O}(-1)$.

Proof: Lemma 2,4) gives $2b-e = 2-2g$.
Since $b>e$, we get $2(1-g) > e$. First suppose $e<0$. Then one knows (Nagata [3]) $e \geq -g$, hence $2(1-g) > -g$, i.e. $g<2$. If $g = 1$, then $e = -1$, but $e \equiv 0(2)$, contradiction. $g = 0$ is not possible if $e<0$.
So $e \geq 0$. Then $2(1-g)>0$, i.e. $g = 0$ and furthermore $e = 0$.
So $N_{S|X} = \mathcal{O}(a) \oplus \mathcal{O}(a)$ for some a.
Since $c_1(N_{S|X}) = -2$ (adjunction formula), $a = -1$.

Remarks: $c_1(K_X|S) = 0$ is satisfied if
a) $S \sim 0$ (homology) or b) $\tilde{S}$ defines an extremal ray on $\hat{X}$ (in the sense of Mori [2]).

Proof of b) : If $R = \mathbb{R}_+[\tilde{S}]$ is an extremal ray, let $\phi : \hat{X} \longrightarrow Y$ be the associated contraction ([2]). Clearly R is not nef ([2]), so ϕ is a modification with exceptional divisor D and $\tilde{S} \subset D$. Since ϕ contracts exactly those curves which are homologous to a positive rational multiple of $\tilde{S}$, D cannot be $\mathbb{P}_2$ or a singular quadric in $\mathbb{P}_3$, so D must be ruled. Necessarily $D = \hat{S}$. So $\hat{S} = \mathbb{P}_1 \times \mathbb{P}_1$ (because Y is projective, so D must have two rulings) and

$$N_{S|X} = \mathcal{O}(-1) \oplus \mathcal{O}(-1).$$

Theorem 4:

The case $c_1(K_X|S) < 0$ is impossible.

Proof: Assume that $c_1(K_X|S) < 0$.
We deduce from lemma 2:

$$2(1-g) + c_1(K_X|S) > e,$$

hence $2(1-g) > e$, (*). Suppose $e<0$. Then by $e \geq -g$ we get $g<1$ which is impossible.
So $e \geq 0$. Then (*) implies $g = 0$. But if $S = \mathbb{P}_1$ and $c_1(K_X|S) < 0$, S can be deformed: denoting by T_X the tangent bundle of X and applying Riemann-Roch, we get $\chi(S,T_X|S) = c_1(K_X^{-1}|S) + 3 > 3$, hence S can be deformed. But clearly S must be rigid, contradiction.

Remarks: It remains to treat the case $c_1(K_X|S)>0$. In this situations it seems much more difficult to get informations. At present I know only the following.
1) If $e \geq 0$, then $N_{S|X}$ is negative and hence S exceptional.
2) If $c_1(K_{\hat{X}}|\tilde{S}) < 0$, then $g = 0$.

Proof: We have the following formula:

$$c_1(K_{\hat{X}}|\tilde{S}) = 2g - 2 + b .$$

So in our situation: $b < 2g - 2$. If $e<0$, then (because $b>e\geq-g$) : $2 - 2g > -g$, hence $g<2$, i.e. $g = 1$ and $e = -1$, $b\geq 0$, contradicting $b<2g-2$. If $e\geq 0$, then $2-2g>b>0$, hence $g = 0$. (Furthermore: $b = 1$, $e = 0$ and $N_{S|X} = \mathcal{O}(-1)\oplus \mathcal{O}(-1)$).

References

[1] Hartshorne, R.: Algebraic Geometry. Graduate Texts in Math. 52, Springer, Berlin-Heidelberg-New-York 1977

[2] Mori, S.: Threefolds whose canonical bundles are not numerically effective. Ann. Math. 116, 133 - 176

[3] Nagata, M.: On self intersection number of a section on a ruled surface. Nagoya Math. J. 37, 191 - 196 (1970)

[4] Peternell, T.: A rigidity theoren for $\mathbb{P}_3(\mathbb{C})$. manuscr. math. 50, 397 - 428 (1985)

[5] Peternell, T.: Algebraic structures on certain 3-folds. Preprint, Bayreuth (1985)

[6] Peternell, T.: Algebraicity criteria for compact complex manifolds. Preprint, Münster 1984/85

On the structure of 4 folds with a hyperplane section which is a $\mathbb{P}^1$ bundle over a ruled surface

by

Eiichi Sato

and

Heinz Spindler

Introduction.

This short note is a supplement to a joint work of the first author with L. Fania and A. Sommese [1]. We prove the following

<u>Theorem 1.</u> Let V be a rank two vector bundle on a smooth complete curve C and $\tau : S = \mathbb{P}(V) \longrightarrow C$ the associated geometrically ruled surface.
If A is an ample divisor on a 4-dimensional projective variety which is a local complete intersection and if A is also a $\mathbb{P}^1$ bundle $\pi : A \longrightarrow S$ over S, then π can be extended to a $\mathbb{P}^2$ bundle $\bar{\pi} : X \longrightarrow S$.

This theorem is already proved in [1] if V is not stable. The key point in dealing with the stable case is the fact that symmetric powers of semistable bundles over curves remain semistable.
The main theorem (1.0) of [1] can now be stated as follows

<u>Theorem.</u> Let X be a 4-dimensional projective variety which is a

local complete intersection. Let A be an ample divisor on X which is a $\mathbb{P}^1$ bundle $\pi : A \longrightarrow S$ over a smooth surface S. Assume that there is a surjective holomorphic map form S to a curve C. Then π can be extended to a $\mathbb{P}^2$ bundle $\bar{\pi} : X \longrightarrow S$.

We would like to express our thanks to the Max-Planck-Institut für Mathematik for making this joint work possible.

1.

The first important step in proving theorem 1 was already done in [1]: Let A be an ample divisor on X, $\tau : S = \mathbb{P}(V) \longrightarrow C$ the ruled surface over the smooth complete curve C and $\pi : A \longrightarrow S$ a $\mathbb{P}^1$ bundle. We extend $\tau\pi : A \longrightarrow C$ to a holomorphic map $\psi : X \longrightarrow C$ [5]. For general $c \in C$ the fibre $A_c = (\tau\pi)^{-1}(c)$ is an ample divisor in $X_c = \psi^{-1}(c)$ and A_c is a $\mathbb{P}^1$ bundle over $S_c = \tau^{-1}(c) \cong \mathbb{P}^1$.
The critical case occurs if $A_c \cong \mathbb{P}^1 \times \mathbb{P}^1$. Then we have the two projections $\pi_c = \pi/A_c : A_c \longrightarrow S_c \cong \mathbb{P}^1$ and $q_c : A_c \longrightarrow \mathbb{P}^1$ and either α) π_c or β) q_c can be extended to a $\mathbb{P}^2$ bundle $X_c \longrightarrow \mathbb{P}^1$ such that $\mathcal{O}_{X_c}(A_c)$ is the universal line bundle on X_c.
By the arguments in [1] we are ready if we can exclude case β).
So let us assume β). Then we get the following situation [1]: There is a second geometrically ruled surface $\tau' : S' = \mathbb{P}(V') \longrightarrow C$ over C and a second projection $\pi' : A \longrightarrow S'$ such that

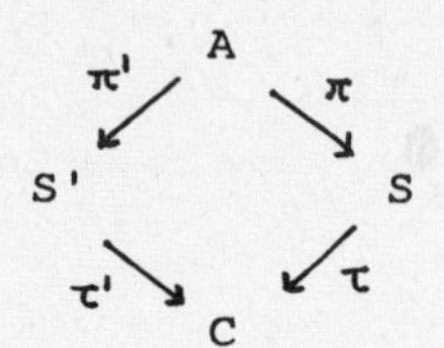

is a cartesian diagram. There is moreover an extension

$$(1) \qquad 0 \longrightarrow \mathcal{O}_{S'} \longrightarrow E \longrightarrow F \longrightarrow 0$$

such that $A \subset X$ coincides with the embedding $\mathbb{P}(F) \subset \mathbb{P}(E)$ induced by (1), where $F = \tau'^*V \otimes \mathcal{O}_{S'}(a)$ for some integer $a \geq 2$.
The ampleness of the divisor A on X exactly means that E is an

ample vector bundle on S'.
Therefore the sequence (1) cannot split and this implies

$$H^1(S', F^\vee) \neq 0. \tag{2}$$

On the other hand using the Kodaira vanishing theorem it was proved in [1] that $H^1(S',F^\vee) = 0$ if the bundle V is not stable.
To prove that ß) cannot occur we only need to show that $H^1(S',F^\vee)$ vanishes in the stable case too. This is the content of the next section.

2.
As we saw in the last section the proof of theorem 1 is reduced to the proof of the following

Theorem 2. Let C be a smooth complete curve and V, V' rank-2 bundles over C. We assume that V is semistable and consider the bundle $F = \tau'^*V \otimes \mathcal{O}_{S'}(a)$ on the projective bundle $\tau' : S' = \mathbb{P}(V') \to C$ associated with V', where a is some integer $a \geq 2$. If F is ample on S' then $H^1(S',F^\vee) = 0$.

Proof. We show: If $H^1(S', F^\vee) \neq 0$ then F is not ample. By Serre duality and the Leray spectral sequence we have (cf. [3] p. 253)

$$H^1(S', F^\vee) \cong H^0(C, V^\vee \otimes S^{a-2} V'^\vee \otimes \det V'^\vee)$$

and so from our assumption $H^1(S', F^\vee) \neq 0$ we get a non zero homomorphism

$$\varphi : S^{a-2} V' \longrightarrow V^\vee \otimes \det V'^\vee . \tag{3}$$

a) First assume that V' is semistable. By a result of Gieseker [2] $S^{a-2}V'$ is semistable too. Since also V is semistable we obtain

$$\frac{\deg S^{a-2}V'}{\operatorname{rk} S^{a-2}V'} \leq \frac{\deg(V^\vee \otimes \det V'^\vee)}{2} .$$

Evaluation of this inequality gives us

$$(a-2) \deg V' \leq - \deg V - 2 \deg V'$$

and therefore

(4) $$c_2(F) = a \deg V + a^2 \deg V' \leq 0$$

Thus F cannot be ample [4].

b) Now we assume that V' is not stable. Then we get an extension

$$0 \to L_1 \to V' \to L_2 \to 0$$

with $\deg L_1 \geq \deg L_2$ and a filtration

$$0 \subset F_o \subset F_1 \subset \ldots \subset F_{a-2} = S^{a-2}V'$$

with

$$F_i = S^i V' \otimes L_1^{a-2-i} \quad , \quad F_i/F_{i-1} = L_2^i \otimes L_1^{a-2-i}$$

for $i = 0, \ldots, a-2$.
Since the homomorphism (3) is non zero, there is some i such that

$$\varphi | F_{i-1} = 0 \ , \ \varphi | F_i \neq 0 \ .$$

So we obtain a non zero homomorphism

$$\tilde{\varphi} : F_i/F_{i-1} \longrightarrow V^\vee \otimes \det V'^\vee \ .$$

By semistability of V we get

(5) $$2 \deg F_i/F_{i-1} \leq - \deg V - 2(\deg L_1 + \deg L_2).$$

Using $\deg L_1 \geq \deg L_2$ we get

(6) $$2 \deg F_i/F_{i-1} = 2(i(\deg L_2 - \deg L_1) + (a-2) \deg L_1)$$
$$\geq 2(a-2) \deg L_2 \ .$$

Combining (5) and (6) we have

(7) $$\deg V + 2(a-1) \deg L_2 + 2\deg L_1 \leq 0 \ .$$

The invertible quotient $V' \twoheadrightarrow L_2$ on the other ahnd gives us a section

$$s : C \longrightarrow S' = \mathbb{P}(V')$$

with $s^* \mathcal{O}_{S'}(1) = L_2$. Therefore we find

$$s^*F = V \otimes L_2^a .$$

From $\deg L_1 \geq \deg L_2$ and (7) we now obtain

$$\begin{aligned} \deg s^*F &= \deg V + 2a \deg L_2 \\ &\leq \deg V + 2(a-1) \deg L_1 + 2\deg L_1 \leq 0. \end{aligned}$$

Again F cannot be ample. The Theorem is proven.

Remark. Theorem 2 holds true under the weaker assumption that rk V and rk V' are arbitrary and without assuming the semistability of V .

References

[1] Fania, M.L., Sato, E., Sommese A.J.: On the strucure of 4 folds with a hyperplane section which is a $\mathbb{P}^1$ bundle over a surface that fibres over a curve.
To appear in Nagoya Math. J. 1985

[2] Gieseker, D.: On a Theorem of Bogomolov on Chern Classes of Stable Bundles. Amer. J. Math. 101, 77-85 (1979)

[3] Hartshorne, R.: Algebraic Geometry. GTM 52, Springer 1977

[4] Kleiman, S.L.: Ample vector bundles on surfaces. Proc. Amer. Math. Soc. 21 (1969), 673-676

[5] Sommese, A.J.: On manifolds that cannot be ample divisors, Math. Ann. 221 (1976), 55-72.

Max-Planck-Institut für Mathematik
Gottfried-Claren-Str. 26
5300 Bonn 3

Sonderforschungsbereich 170
Geometrie und Analysis
Bunsenstr. 3-5
3400 Göttingen

COMPLEX SURFACES WITH NEGATIVE TANGENT BUNDLE

Michael Schneider

INTRODUCTION

In contrast to compact Riemann surfaces the uniformization of projective algebraic manifolds of higher dimension is still very little understood. In this connection S. Yau [21] has raised the question to determine all projective surfaces having negative tangent bundle.

B. Wong [20] and F. Bogomolov have constructed simply connected surfaces with negative tangent bundle. This (almost) contradicts one of Yau's conjectures in [21].

In [16] a "numerical" criterion for the negativity of a rank 2 vector bundle on a projective surface was given. For the tangent bundle T_X this reads:

i) $c_1(X)^2 > 2c_2(X) > 0$

ii) $T_X|C$ is negative for all integral curves $C \subset X$

imply that T_X is negative.
As promised in [16] this will be applied here to check the negativity of the tangent bundle of Kodaira surfaces [11]. As I have been informed in the meantime this has been established by M. Deschamps [1] in a different way.

Submanifolds X of simple abelian varieties A have been suspected to have negative tangent bundle if $\dim X \leq \operatorname{codim} X$. We will construct an example showing that this is not true in general.

Finally we consider the problem of finding all ratios $c_1(X)^2/c_2(X)$ of Chern numbers of surfaces with negative tangent bundle. For surfaces of general type the determination of the Chern numbers c_1^2 and c_2 has found much attention [6],[7],[8],[12],[13],[15],[17],[19].

It is a pleasure to thank R. Coleman, R. Lazarsfeld, A. Sommese and H. Wu for stimulating discussions and H. Spindler for pointing out the paper [1].

PRELEMINARIES

A holomorphic vector bundle E on a compact complex space is negative (in the sense of Grauert) if the zero section of E admits a strongly pseudoconvex neighborhood. This is equivalent [3] to the possibility of contracting the zero section of E analytically to a point. The negativity of E is equivalent to the ampleness of the dual bundle E^* in the sense of Hartshorne [4], i.e. $\mathcal{O}_{\mathbb{P}(E^*)}(1)$ is an ample line bundle on $\mathbb{P}(E^*)$, the projective bundle of hyperplanes in the fibres of E^*. The Chern classes of a vector bundle E will be denoted by $c_i(E)$. For the holomorphic tangent bundle T_X we write $c_i(X) = c_i(T_X)$. In case X is a surface, $c_1(X)^2$ and $c_2(X)$ will be considered as integers.
We will not distinguish between vector bundles and locally free sheaves.

We collect now some elementary facts which will be needed later.

LEMMA 1. *Let S be a compact Riemann surface and let*

$$0 \to E \to F \to \mathcal{O} \to 0$$

be an exact sequence of vector bundles which does not split.
If E is negative then F is negative too.
For a proof we refer to [2].

Let $\pi : X \to S$ be a surjective smooth holomorphic map of compact connected complex manifolds.
The sequence of tangent bundles

$$(*) \qquad 0 \to T_{X/S} \to T_X \to \pi^*T_S \to 0$$

leads to an exact sequence

$$0 \to \pi_*T_{X/S} \to \pi_*T_X \to T_S \xrightarrow{\delta} R^1\pi_*T_{X/S} \to \cdots$$

The connecting homomorphism

$$\delta : T_S \to R^1\pi_*T_{X/S}$$

is the Kodaira-Spencer map.
Recall that $\delta = 0$ means that π is locally trivial. At a point $s \in S$ one gets

$$\delta(s) : T_{S,s} \to H^1(X(s), T_{X(s)}) .$$

LEMMA 2. *Suppose* $H^o(X(s), T_{X(s)}) = 0$ *for all* $s \in S$.
For $s \in S$ *we consider the restriction of (*) to* $X(s)$:

$$(*)_s \qquad 0 \to T_{X(s)} \to T_X|X(s) \to (\pi^* T_S)|X(s) \to 0.$$

Then

i) *If* $(*)_s$ *splits,* $\dim H^o(X(s), T_X|X(s)) = \dim S$

ii) $\dim H^o(X(s), T_X|X(s)) = \dim S$ *is equivalent to* $\delta(s) = 0$.

iii) *For* $\dim S = 1$ *the splitting of* $(*)_s$ *is equivalent to* $\dim H^o(X(s), T_X|X(s)) = \dim S$.

The proof is left to the reader.

We will use Lemma 2 in the case that S is a Riemann surface and X a complex surface. If the fibres have genus at least 2 the hypothesis of Lemma 2 is satisfied. Note also, that in this case $\delta \neq 0$ implies that the genus of S and the genus of the fibres is at least 2.

COROLLARY. *Let* $\pi : X \to S$ *be a smooth surjective map of the compact connected complex surface onto the compact Riemann surface* S. *Suppose* $\delta(s) \neq 0$ *for all* $s \in S$. *Then* $T_X|X(s)$ *is negative for all* $s \in S$.

Proof. Consider the sequence $(*)_s$

$$0 \to T_{X(s)} \to T_X|X(s) \to \mathcal{O}_{X(s)} \to 0.$$

Since $\delta \neq 0$ we can apply Lemma 2 to see that $(*)_s$ does not split for any $s \in S$. Since $T_{X(s)}$ is negative, Lemma 1 implies the negativity of $T_X|X(s)$.

SOME SURFACES WITH NEGATIVE TANGENT BUNDLE.

In this section we will show the negativity of the tangent bundle of a class of surfaces which includes the Kodaira surfaces [11],[9].

THEOREM 1. *Let* $\pi : X \to S$ *be a smooth holomorphic map of a connected complex projective surface onto a compact Riemann surface. Assume* $\delta(s) \neq 0$ *for all* $s \in S$. *Then* T_X *is negative.*

Proof. By [16] it is enough to show that

i) $c_1(X)^2 > 2c_2(X) > 0$

ii) $T_X|C$ is negative for all integral curves $C \subset X$.

In order to prove i) consider the sequence

$$(*) \qquad 0 \to T_{X/S} \to T_X \to \pi^* T_S \to 0 .$$

Since $\delta \neq 0$ by assumption we can apply Paršin's theorem [14](see also [18]):

$$T_{X/S} \text{ is negative.}$$

In particular $c_1(T_{X/S})^2 > 0$. But from (*) it is immediate that

$$c_1(X)^2 - 2c_2(X) = c_1(T_{X/S})^2 .$$

From (*) we have also

$$\begin{aligned} c_2(X) &= c_1(T_{X/S})\pi^*(c_1(S)) \\ &= c_1(\Omega_{X/S})\pi^*(c_1(\Omega_S)). \end{aligned}$$

Since S is of genus at least two, Ω_S is generated by sections and therefore $\pi^*(c_1(\Omega_S))$ is represented by a multiple of a fibre. Since $\Omega_{X/S}$ is ample by Paršin we obtain the positivity of $c_2(X)$.

To establish ii) let $C \subset X$ be an integral curve. Two cases are possible:

a) $\pi(C) = S$.

Then $f = \pi|C$ is finite and therefore $f^*(T_S)$ is negative (S has genus at least 2). Restrict (*) to C to obtain

$$0 \to T_{X/S}|C \to T_X|C \to f^*T_S \to 0 .$$

Since $T_{X/S}$ is negative it follows that $T_X|C$ is negative.

b) C is a fibre of π.

Then apply the Corollary to Lemma 2 to conclude that $T_X|C$ is negative.

COROLLARY. Kodaira surfaces have negative tangent bundle.

Proof. Kodaira surfaces are of the type considered in Theorem 1, [11], [9]. Kodaira has shown $\delta(s) \neq 0$ for all $s \in S$.

In case one assumes in Theorem 1 only $\delta \neq 0$ one can still show

THEOREM 2. Let $\pi : X \to S$ *be a smooth holomorphic map from the connected projective surface onto the Riemann surface* S. *If* $\delta \neq 0$ *the tangent bundle* T_X *is negative almost everywhere.*

Proof. As in the proof of theorem 1 one shows

$$c_1(X)^2 > 2c_2(X).$$

Moreover the canonical bundle K_X is ample since

$$K_X = \Omega_{X/S} \otimes \pi^* K_S .$$

Therefore X is of general type and we can apply Miyaoka's theorem [12] to conclude that Ω_X is ample almost everywhere. This implies that T_X is negative almost everywhere.

REMARK. Various definitions of almost everywhere ampleness of a vector bundle are possible. Let us recall here Miyaoka's notion: A vector bundle E on the projective manifold X is ample a.e. if there exists a proper Zariski closed subset $T \subset X$ and $\varepsilon > 0$ such that

$$c_1(\mathcal{O}_{\mathbb{P}(E)}(1)|C) > \varepsilon c_1(H|C)$$

for all integral curves $C \subset \mathbb{P}(E)$ whose projection is not contained in T. Here H is an (arbitrary) ample line bundle on $\mathbb{P}(E)$.

SOME SURFACES WHOSE TANGENT BUNDLE IS NOT NEGATIVE.

In this section we will construct a surface in a 4-dimensional simple abelian variety whose tangent bundle is not negative. This contradicts a standard conjecture.

PROPOSITION 1. Let $f : X \to Y$ *be a finite surjective holomorphic map of compact complex manifolds and let* $R = \{x \in X : \text{rank of } f \text{ at } x \text{ is not maximal}\}$. *Suppose that there is a closed analytic subspace* $A \subset X$, $A \not\subset R$ *such that* $H^0(T_X|A) \neq 0$.
Then T_Y *can not be negative.*

Proof. Consider the map

$$T_X \to f^* T_Y ,$$

which is an injection of sheaves. Since $A \not\subset R$ we get an injection

$$T_X|A \to (f^* T_Y)|A$$

and therefore $H^o(T_X|A) \subset H^o((f^*T_Y)|A)$. This shows $H^o((f^*T_Y)|A) \neq 0$, hence $(f^*T_Y)|A$ is not negative. A fortiori f^*T_Y is not negative. Since f is finite T_Y can not be negative either.

COROLLARY 1. Let X_1, X_2 be positive-dimensional compact complex manifolds and let

$$f : X_1 \times X_2 \to Y$$

be a finite holomorphic map onto the complex manifold Y. Then T_Y can not be negative.

Proof. $T_{X_1 \times X_2} = pr_1^*T_{X_1} \oplus pr_2^*T_{X_2}$. Pick $a \in X_1$ such that $A = a \times X_2$ is not contained in $R = \{x \in X_1 \times X_2 : \text{rk} f \text{ is not maximal at } x\}$.

Then $H^o(T_{X_1 \times X_2}|A) \simeq H^o(X_2; T_{X_2} \oplus \mathcal{O}^{\dim X_1}) \neq 0$.

Now apply the proposition.

COROLLARY 2. The symmetric product S^2C of a smooth curve C never has negative tangent bundle.

THEOREM 3. There exist simple abelian varieties A (of dimension at least 4) and smooth surfaces $X \subset A$ such that T_X is not negative.

Proof. Pick a smooth curve C of genus 4, not hyperelliptic, such that the Jacobian $A = Jac(C)$ of C is simple. The possibility of doing this has been pointed out to me by R. Coleman. By Corollary 2 above the tangent bundle of $X = S^2C$ is not negative. But we have an embedding

$$S^2C \to A = Jac(C).$$

REMARKS. 1) By a result of Hartshorne [5] every submanifold of a simple abelian variety has ample normal bundle.
2) It was suspected that submanifolds X of simple abelian varieties satisfying $\dim X \leq \text{codim} X$ have negative tangent bundle.

CHERN NUMBERS OF SURFACES WITH (ALMOST) NEGATIVE TANGENT BUNDLE.

The geography of Chern numbers of surfaces of general type has been studied quite intensively [6],[7],[8],[12],[13[,[15],[17],[19]. The Chern numbers satisfy the estimates $\frac{1}{5}(c_2-36) \leq c_1^2 \leq 3c_2$ and Sommese [17] has shown that all rational numbers in $[\frac{1}{5},3]$ occur as c_1^2/c_2 of a surface of general type.

Surfaces with negative tangent bundle have ample cotangent bundle and are therefore of general type. For an ample vector bundle E of rank 2 on a surface one has always $c_1(E)^2 > c_2(E) > 0$ by Kleimann [10]. Hence for surfaces X with negative tangent bundle the ratios $c_1(X)^2/c_2(X)$ lie in the interval $]1,3]$.

PROBLEM. Which $q \in]1,3] \cap \mathbb{Q}$ occur as $c_1(X)^2/c_2(X)$ of a surface X with negative tangent bundle T_X? Is it true that the ratios $c_1(X)^2/c_2(X)$ are dense in $]1,3]$?

The known methods to construct surfaces with negative tangent bundle yield many values of c_1^2/c_2 but nothing like density. It is not hard to construct examples of surfaces X with negative tangent bundle whose ratio $c_1(X)^2/c_2(X)$ is arbitrarily close to 1.

The above Problem can be weakened by asking which $q \in]1,3] \cap \mathbb{Q}$ occur as ratios $c_1(X)^2/c_2(X)$ of a surface X with almost everywhere negative tangent bundle. Using Sommese's results [17] it is not hard to show the following

THEOREM 4. For $q \in [2,3] \cap \mathbb{Q}$ there exists a projective surface X whose tangent bundle is negative almost everywhere such that

$$c_1(X)^2/c_2(X) = q .$$

REFERENCES

[1] DESCHAMPS-MARTIN, M.: Propriétés de descente des variétés à fibré cotangent ample. Ann. Inst. Fourier, Grenoble 33 (1984), 39-64.

[2] GIESEKER, D.: p-ample bundles and their Chern classes. Nagoya Math. J. 43 (1971), 91-116.

[3] GRAUERT, H.: Über Modifikationen und exzeptionelle analytische Mengen. Math. Ann. 146 (1962), 331-368.

[4] HARTSHORNE, R.: Ample vector bundles. Publ. Math. IHES 29 (1966), 63-94.

[5] HARTSHORNE, R.: Ample vector bundles on curves. Nagoya Math. J. 43 (1971), 73-89.

[6] HIRZEBRUCH, F.: Arrangements of lines and algebraic surfaces. Arithmetic and geometry, Vol. II. Progress in Math. Vol. 36, 113-140. Boston, Basel, Stuttgart: Birkhäuser 1983.

[7] HOLZAPFEL, R.P.: A class of minimal surfaces in the unknown region of surface geography. Math. Nachr. 98 (1980), 211-232.

[8] HOLZAPFEL, R.P.: Invariants of arithmetic ball quotient surfaces. Math. Nachr. 103 (1981), 117-153.

[9] KAS, A.: On deformations of a certain type of irregular algebraic surface. Amer. J. Math. 90 (1968), 789-804.

[10] KLEIMAN, S.: Ample vector bundles on surfaces. Proc. Amer. Math. Soc. (1969), 673-676.

[11] KODAIRA, K.: A certain type of irregular algebraic surfaces. Journal d'Analyse Math. 19 (1967), 207-215.

[12] MIYAOKA, Y.: Algebraic surfaces with positive indices. Classification of algebraic and analytic manifolds. Katata Symposium Proc. 1982. Progress in Math. vol. 39, 281-301. Boston, Basel, Stuttgart: Birkhäuser 1983.

[13] MOSTOW, G.D., SIU, Y.T.: A compact Kähler surface of negative curvature not covered by the ball. Ann. of Math. 112 (1980), 321-360.

[14] PARSIN, A.N.: Algebraic curves over function fields I. Math. USSR Izvestija 2 (1968), 1145-1170.

[15] PERSSON, U.: Chern invariants of surfaces of general type. Comp. Math. 43 (1981), 3-58.

[16] SCHNEIDER, M., TANCREDI, A.: Positive vector bundles on complex surfaces. Manuscripta math. 50 (1985), 133-144.

[17] SOMMESE, A.: On the density of ratios of Chern numbers of algebraic surfaces. Math. Ann. 268 (1984), 207-221.

[18] SZPIRO, L.: Propriétés numériques du faisceau dualisant relatif. Séminaire sur les pinceaux de courbes de genre an moins 2. Exposé n° 3, Astérisque n° 86 (1981), 44-78.

[19] VAN DE VEN, A.: Some recent results on surfaces of general type. Sém. Bourbaki 1976/77 exposé 500. Lecture Notes in Mathematics 677 (1978), 155-166. Springer.

[20] WONG, B.: A class of compact complex manifolds with negative tangent bundles. AMS Proc. of Symposia in Pure Math. 41 (1984), 217-223.

[21] YAU, S.T.: Seminar on Differential geometry. Annals of Math. Studies 102. Princeton Univ. Press 1982.

Michael Schneider
Mathematisches Institut
der Universität
Postfach 3008
D-8580 Bayreuth
West-Germany

NONEQUIDIMENSIONAL VALUE DISTRIBUTION THEORY AND SUBVARIETY EXTENSION

Yum-Tong Siu [1]

§1 Statement of Result on Defect Relation

In this talk we present a result, with a sketch of its proof, on the defect relation for the value distribution of holomorphic maps between manifolds of *unequal* dimensions. A complete proof will be published elsewhere. And we discuss the possibility of developing a theory of essential singularities for complex-analytic subvarieties which corresponds to a local version of the value distribution theory for holomorphic functions and maps. The result on defect relation given here can be regarded as a first step toward the development of such a local value distribution theory for holomorphic maps.

For the sake of simplicity we will deal with the simple, yet typical, case of a holomorphic map from $\mathbb{C}$ to a compact complex surface and then merely indicate the corresponding result for the higher dimensional case. The defect relation for a holomorphic map from $\mathbb{C}$ to a compact complex surface is given as follows.

Theorem 1. Suppose M is a compact complex manifold of complex dimension two and L, F are positive holomorphic line bundles over M. Suppose s_j $(j \in J)$ is a holomorphic section of L over M with nonsingular zero-set so that no more than k of the zero-sets of s_j have a common intersection point. Suppose $\Gamma^{\gamma}_{\alpha\beta}$ is a meromorphic connection for the tangent bundle of M such that for some nonzero holomorphic section t of F, $t\Gamma^{\gamma}_{\alpha\beta}$ is holomorphic. Assume that locally on M for any fixed coordinate indices α, β of M and any fixed j, $t\mathfrak{D}_{\alpha}\partial_{\beta}s_j$ is a linear combination of $\partial_{\alpha}s_j$, $\partial_{\beta}s_j$, and s with smooth coefficients, where ∂_{α} is the $(1,0)$ partial differentiation with respect to the αth holomorphic coordinate of M (after a local trivialization of L) and $\mathfrak{D}_{\alpha}$ is the $(1,0)$ covariant differentiation with respect to the connection $\Gamma^{\gamma}_{\alpha\beta}$. Let f be a holomorphic map from $\mathbb{C}$ to M whose characteristic function $\mathcal{T}_r(f^*\theta_L)$

---

[1] Research partially supported by a National Science Foundation grant.

with respect to the curvature form θ_L of L grows faster than the logarithmic order $\log r$ of the Euclidean distance function r of $\mathbb{C}$ from the origin (that is, $\log r = o(\mathcal{T}_r(f^*\theta_L))$). Suppose the map from $\mathbb{C}$ to the anticanoncial line bundle K_M^{-1} of M defined by the wedge product $f_z \wedge \mathfrak{D}_z f_z$ of the differential f_z of f and its covariant derivative $\mathfrak{D}_z f_z$ with respect to the connection $\Gamma^{\gamma}_{\alpha\beta}$ and the coordinate z of $\mathbb{C}$ is not identically zero. Then the sum of the defects $\delta_f(Zs_j)$ of f for the zero-set Zs_j of s_j $(j \in J)$ does not exceed $k\sigma$ for any nonnegative number σ satisfying the condition that σ times the curvature θ_L of L is no less than the curvature $\theta_{F\otimes K_M^{-1}}$ of $F\otimes K_M^{-1}$ at every point of M. Here

$$\mathcal{T}_r(f^*\theta_L) = \int_{\rho=0}^{r} \frac{d\rho}{\rho} \int_{|z|<\rho} f^*\theta_L$$

and

$$\delta_f(Zs_j) = \liminf_{r\to\infty} \left(1 - \frac{N_f(Zs_j)}{\mathcal{T}_r(f^*\theta_L)}\right)$$

with

$$N_f(Zs_j) = \int_{\rho=0}^{r} (\text{Cardinality of } f^{-1}Zs_j \text{ in } |z| < \rho) \frac{d\rho}{\rho}.$$

Remarks. (1) The nondegeneracy condition on f of the non-identical vanishing of $f_z \wedge \mathfrak{D}_z f_z$ geometrically means that the image of f is not autoparallel with respect to the connection $\Gamma^{\gamma}_{\alpha\beta}$. (2) The condition that locally on M for any fixed coordinate indices α, β of M and any fixed j, $t\mathfrak{D}_\alpha \partial_\beta s_j$ is a linear combination of $\partial_\alpha s_j$, $\partial_\beta s_j$, and s with smooth coefficients geometrically means that "the second fundamental form" of the divisor Zs_j is zero with respect to the connection $\Gamma^{\gamma}_{\alpha\beta}$.

(3) The nonautoparallelism of the image of the map and the vanishing of the second fundamental form of the divisors make this result (in the case of higher dimensional M) a very logical extension, to the case of a general manifold and general divisors, of the result of Weyl-Weyl [18] and Ahlfors [1] on the defect relation for holomorphic maps from $\mathbb{C}$ to the projective space $\mathbb{P}_n$ with hyperplanes as divisors. The relation between our result and the result of Weyl-Weyl and Ahlfors

becomes even clearer in the following version of Theorem 1 for the case of a smooth connection.

Theorem 2. Suppose in Theorem 1 the connection $\Gamma^{\gamma}_{\alpha\beta}$ is smooth instead of meromorphic. If $f_z \wedge \mathfrak{D}_z f_z$ is holomorphic, then the sum of the defects $\delta_f(Zs_j)$ does not exceed $k\sigma$ when σ satisfies $\sigma\,\theta_L \geq \theta_{K_M^{-1}}$ everywhere on M.

Remark. When $M = \mathbb{P}_2$ and $\Gamma^{\gamma}_{\alpha\beta}$ is the Levi-Civita connection for the Fubini-Study metric, as a consequence of the constancy of the holomorphic sectional curvature of the projective space, $f_z \wedge \mathfrak{D}_z f_z$ is always holomorphic for any holomorphic map f from $\mathbb{C}$ to M. In a certain sense this explains why one gets the sharp defect relation of Weyl-Weyl and Ahlfors for maps from $\mathbb{C}$ to $\mathbb{P}_2$. (Analogous situations hold for $\mathbb{P}_n$.) In Theorems 1 and 2 when the divisors intersect normally (in particular $k = 2$) the sum of the defect should simply be σ instead of 2σ and this has not yet been verified.

Example. Let w_0, w_1, w_2 be the homogeneous coordinates of $\mathbb{P}_2$ and $\zeta_\nu = \frac{w_\nu}{w_0}$ $(\nu = 1, 2)$ be the inhomogeneous coordinates. Let $M = \mathbb{P}_2$ and $L = H^d$ (where H is the hyperplane section line bundle of $\mathbb{P}_n$) and s_j be the holomorphic section $\Sigma^2_{\nu=0} a^{(j)}_\nu w_\nu$ for some nonzero $(a^{(j)}_0, a^{(j)}_1, a^{(j)}_2) \in \mathbb{C}^3$. The divisor Zs_j is a Fermat curve of degree d. In this case we can choose the holomorphic line bundle F to be H^3 and the connection $\Gamma^{\gamma}_{\alpha\beta}$ in the inhomogeneous coordinates ζ_1, ζ_2 to be given by $\Gamma^{\nu}_{\nu\nu} = -\frac{d-1}{\zeta_\nu}$ $(\nu = 1, 2)$ with all the other components of $\Gamma^{\gamma}_{\alpha\beta}$ being zero. According to Theorem 1, the sum of the defects of a holomorophic map $f:\mathbb{C} \to M$ for the divisors Zs_j does not exceed $\frac{6k}{d}$ if no more than k of the divisors Zs_j have a common point of intersection and f satisfies the nondegeneracy condition. This defect relation agrees with the following result of M. Green [7] in the situation common to both our result and his. A holomorphic map from $\mathbb{C}^m$ to $\mathbb{P}_n$ whose image omits the Fermat variety of degree $d > n(n+1)$ must have its image contained in a linear subspace of dimension $\leq \frac{n}{2}$

and examples show that the upper bound $\frac{n}{2}$ for the dimension of the linear subspace is sharp. In particular, under Green's nondegeneracy condition, for $n = 2$ the degree d cannot be bigger than 6. This number agrees with the number from our result on the defect relation in the case $k = 1$ because if $d > 6$ then the defect $\frac{6k}{d}$ is < 1 and the image of the map cannot omit completely the Fermat variety.

§2. Possibility of Local Value Distribution Theory.

Before we sketch the proof of the above results on the defect relation, we would like to discuss the possibility of developing a theory of essential singularities for complex-analytic subvarieties which corresponds to a local version of the value distribution theory for holomorphic functions and maps.

The classical value distribution theory of Nevanlinna can be regarded as the theory about the behavior of a holomorphic funciton from $\mathbb{C}$ to $\mathbb{C}$ near its essential singularity at infinity. The precursors of the theory of Nevanlinna are the theorems of Riemann and Picard asserting respectively that the image of such a function is dense and that the image cannot miss two points. The theory of Nevanlinna tells us that almost every point in the image space is assumed an equal number of times determined by the growth of the function when appropriately counted. Moreover, the sum of the defect at all points of the image space which measures the failure of the point to be assumed the expected number of times does not exceed two.

This theory of Nevanlinna was generalized by H. Weyl, J. Weyl [18] and Ahlfors [1] to holomorphic maps from $\mathbb{C}$ to the complex projective n-space and by Stoll [15,16], Griffiths [4,5], M. Green [7], Shiffman [10,11], and others to the equidimensional case of holomorphic maps from $\mathbb{C}^n$ to a projective algebraic manifold of complex dimension n and some other situations. For analytic objects other than holomorphic maps one has some cruder results about their essential singularities. More closely related to holomorphic maps than other analytic objects is the complex-analytic subvariety, because the graph of a holomorphic map is a subvariety. For the theory of essential singularities of subvarieties, for example, we have the theorem of Thullen-Remmert-Stein [9,17] and the theorem of Stoll-Bishop [3,14]. Having an essential singularity at one point simply means that there is no analytic extension at that point.

The theorem of Thullen (with the higher dimensional generalization due to Remmert-Stein) corresponds to the theorem of Riemann on the density of the image of a holomorphic function with essential singularity. To put it in the context of essential singularities one can state it in the following way. Let Z be an irreducible complex-analytic curve in a domain Ω in $\mathbb{C}^2$ and V be an irreducible complex-analytic curve in $\Omega - Z$. One says that V has an essential singularity at Z if V cannot be extended to a complex-analytic curve in Ω (which is equivalent to saying that the topological closure V^- of V in Ω is not complex-analytic in Ω). Thullen's theorem says that if V has an essential singularity at Z, then V is dense at every point of Z in the sense that Z is contained in V^-.

The theorem of Stoll-Bishop (with an earlier version of subvarieties in the Euclidean space due to Stoll) is the following. Let Z be a complex-analytic subvariety in a domain Ω in $\mathbb{C}^n$ and V be a complex-analytic subvariety of pure dimension in $\Omega - Z$. If V has an essential singularity at Z, then the volume of V cannot be finite. This theorem corresponds to the theorem of Riemann on removable singularities, because if Ω is $\Delta \times \mathbb{C}$ (where Δ is the open unit 1-disc) and Z is $0 \times \mathbb{C}$ and V is the graph of a holomorphic function f on $\Delta - 0$. then the volume of V is finite if and only if the derivative of f is square integrable near 0.

There is some indication that there is a possibility of developing a theory of essential singularities for complex-analytic subvarieties analogous to the value distribution theory for holomorphic functions and maps. Since the graph of a holomorphic function or map is the special case of a global subvariety biholomorphic to one factor of the ambient product space, such a theory of essential singularities can be regarded as a local version of the value distribution theory of holomorphic maps and functions.

The value distribution theory of holomorphic functions and maps is based on Stokes' theorem, the concavity of the logarithmic function, estimates of a function by its integral outside a set of finite logarithmic measure, and the use of curvature forms. Until recently the investigation of essential singularities of subvarieties is based on entirely different techniques. For example, the original Bishop's proof of the theorem of Stoll-Bishop involves some ingenious use of

Jensen measures, the Poisson kernel, and volume estimates by geometric measure theory. The value distribution theory and the theory of essential singularities of subvarities seem to be too different to have enough common ground to be combined into a value distribution theory for subvarieties. However, the recent work of Skoda [13], El Mir [6], and Sibony [12] gives a new proof of the theorem of Stoll-Bishop by using Stokes' theorem. It is thus entirely likely that a quantitative theory of essential singularities for subvarieties may be developed in a way analogous to the value distribution theory for holomorphic functions and maps.

Let Z be an irreducible complex-analytic curve in a domain Ω in $\mathbb{C}^2$ and V be an irreducible complex-analytic curve in $\Omega - Z$. With the case of the graph of a holomorphic function as a guide, if V has an essential singularity at Z, then to formulate the statement that V frequents almost every point of Z the same number of times with a defect estimate, one way is to introduce a family of complex-analytic curves C_t transversal to Z and determine how often V intersects a particular C_t when the counting is suitably weighted according to the distance of the point of intersection from Z. To deal with this problem one is compelled to consider the easier question of value distribution theory for a holomophic map into a compact complex manifold of complex dimension two and determine how often the image hits a divisor in the target manifold. This is the question of value distribution theory where the dimensions of the domain and target manifolds are unequal. Besides the theory of holomorphic maps from $\mathbb{C}$ to the projective n-space $\mathbb{P}_n$ with the divisors being hyperplanes due to Weyl-Weyl [18] and Ahlfors [1] (where the linear structure of the projective space and the hyperplanes are used in a crucial way) and some very intimately related results, only very limited information has been obtained about the nonequidimensional value distribution theory [2,5,7,8,11,16]. The possibility of developing a theory of essential singularities for complex-analytic subvarieties which corresponds to a local version of the value distribution theory for holomorphic functions and maps is one of the motivations for investigating the kind of defect relations given in Theorems 1 and 2 in the value distribution theory of holomorphic maps from $\mathbb{C}$ to a compact complex manifold of complex dimension two.

In the case of a curve Z in a domain Ω in $\mathbb{C}^2$ and another curve V in $\Omega - Z$ having an essential singularity at Z, another possibility of formulating the statement that V frequents almost every point of Z the same number of times with a defect estimate is to use the analog of Lelong numbers for V at points of Z.

§3. Sketch of Proof of Defect Relation.

We now sketch the proof of Theorem 1. For notational simplicity we do only the case $k = 1$. To make the sketch of the proof of our theorem easier to understand, we first look at the classical case of value distribution from $\mathbb{C}$ to a compact Riemann surface. (Though there only the case of the projective line $\mathbb{P}_1$ is interesting, in order to see how one deals with the transiton to the higher dimensional case we consider the case of a general compact Riemann surface.) One way to look at value distribution theory is to regard it as a generalization, to the case of noncompact base manifold, of the theorem relating the Chern class of a Hermitian holomorphic line bundle over a compact Riemnan surface as the integral of its curvature to the number of zeroes of a holomorphic section.

Let X be a compact Riemann surface and E be a Hermitian holomoprhic line bundle over X and u be a holomorphic section of E over X. Let $|u|$ denote the pointwise norm of u and θ_E denote the curvature form of E and Zu denote the current of type $(1,1)$ defined by integration over the zero-set of u. The Poincaré-Lelong formula yields

$$\frac{i}{2\pi}\partial\bar{\partial}\log|u|^2 = Zu - \theta_E.$$

Since the left-hand side is exact and X is compact, integrating this formula over X gives us the result that the number of zeroes of u is equal to the integral of θ_E over X which is the first Chern class $c_1(E)$ of E.

For value distribution theory for a holomorphic map f from $\mathbb{C}$ to a compact Riemann surface M with a Hermitian holomorphic line bundle L and a holomorphic section s, we replace X by $\mathbb{C}$ and replace E by the pullback f^*L of L under f and replace u by the pullback f^*s of s under f. Integration over a disc of finite radius in $\mathbb{C}$ now yields the following equation with a boundary term

$$\int_{\partial B_\rho} \frac{i}{2\pi} \bar\partial \log|f^*s|^2 = \int_{B_\rho} Zf^*s - \int_{B_\rho} f^*\theta_L,$$

where B_ρ is the disc of radius ρ in $\mathbb{C}$ centered at the origin and ∂B_ρ is its boundary. The main problem in value distribution theory is to show that under certain conditions and with suitable interpretations one can conclude that the contribution from the boundary term is negligible and thereby obtain the result that the number of times the image of f hits the zero-set of s is measured by the integral of the pullback of the curvature form of L. To get rid of the differentiation inside the integral of the boundary, one integrates the above equation with respect to $\frac{d\rho}{\rho}$ and gets

$$\mathcal{A}_r(\log|f^*s|^2) - \log|f^*s|^2(0) = \mathcal{I}_r(Zf^*s) - \mathcal{I}_r(f^*\theta_L),$$

where $\mathcal{A}_r(g)$ means $\frac{1}{4\pi r}\int_{\partial B_r} g$ and, for a $(1,1)$-form α,

$$\mathcal{I}_r(\alpha) = \int_{\rho=0}^{r} \frac{d\rho}{\rho} \int_{B_\rho} \alpha.$$

(When g is a function, we will use $\mathcal{I}_r(g)$ to denote $\mathcal{I}_r(g\ \eta)$, where η is the Euclidean Kähler form of $\mathbb{C}$.) Since M is compact, one can multiply the Hermitian metric of L by a small positive number to make $|s| < 1$ on M without affecting the right-hand side of the last equation. So without loss of generality we can assume that $\mathcal{A}_r(\log|f^*s|^2)$ is nonpositive and get

$$\mathcal{I}_r(Zf^*s) \le \mathcal{I}_r(f^*\theta_L) - \log|f^*s|^2(0)$$

which is the first Main Theorem of Nevanlinna. It gives us an estimate of the size of the zero-set of f^*s by the integral of the pullback of the curvature form. The term $\mathcal{I}_r(Zf^*s)$ is known as the *counting function* and the term $\mathcal{I}_r(f^*\theta_L)$ is known as the *characteristic function*. The difficult part of value distribution theory is to get an estimate in the other direction bounding $\mathcal{I}_r(f^*\theta_L)$ in terms of $\mathcal{I}_r(Zf^*s)$. It is equivalent to getting an upper estimate of the *defect* $\delta_f(Zs)$ defined by

$$\delta_f(Zs) = \liminf_{r\to\infty} \frac{\mathcal{T}_r(f^*\theta_L) - \mathcal{T}_r(Zf^*s)}{\mathcal{T}_r(f^*\theta_L)}.$$

To get this estimate Nevanlinna introduced the trick of considering $\frac{i}{2\pi}\partial\bar{\partial}\log\frac{1}{|f^*s|^2(\log|f^*s|^2)^2}$ instead of $\frac{i}{2\pi}\partial\bar{\partial}\log|f^*s|^2$. We have

$$\frac{i}{2\pi}\partial\bar{\partial}\log\frac{1}{|f^*s|^2(\log|f^*s|^2)^2} = (f^*\theta_L - Zf^*s)(1 + \frac{2}{\log|f^*s|^2}) + 2\frac{|f^*Ds|^2}{|f^*s|^2(\log|f^*s|^2)^2},$$

where Ds denotes the covariant differential of s. Let C denote the value of $\log\frac{1}{|f^*s|^2(\log|f^*s|^2)^2}$ at the origin. Integrating the above equation twice gives us

$$\begin{aligned}\mathcal{A}_r(\log\frac{1}{|f^*s|^2(\log|f^*s|^2)^2}) - C &= \mathcal{T}_r((f^*\theta_L - Zf^*s)(1 + \frac{2}{\log|f^*s|^2})) \\ &\quad + \mathcal{T}_r(2\frac{|f^*Ds|^2}{|f^*s|^2(\log|f^*s|^2)^2}) \\ &\geq (1-\varepsilon)\mathcal{T}_r(f^*\theta_L) - (1+\varepsilon)\mathcal{T}_r(Zf^*s) + \mathcal{T}_r(2\frac{|f^*Ds|^2}{|f^*s|^2(\log|f^*s|^2)^2}),\end{aligned}$$

when we rescale the Hermitian metric of L to make $\frac{2}{\log|s|^2} \leq \varepsilon$ everywhere on M. This trick yields the additional term

$$\mathcal{T}_r(2\frac{|f^*Ds|^2}{|f^*s|^2(\log|f^*s|^2)^2})$$

at the expense modifying the terms $\mathcal{T}_r(f^*\theta_L)$ and $\mathcal{T}_r(Zf^*s)$ respectively by the factors $1-\varepsilon$ and $1+\varepsilon$. Such modifications by factors $1-\varepsilon$ and $1+\varepsilon$ have no effect in the end because we will eventually let $\varepsilon \to 0$. However, the additional term can be used to help dominate the troublesome boundary term $\mathcal{A}_r(\log\frac{1}{|f^*s|^2(\log|f^*s|^2)^2})$. For this purpose one uses twice the following Calculus Lemma which essentially says that the integrand is dominated by its integral outside a set of controllable measure.

For a positive continuously differentiable function $S(r)$ $(r > 0)$

with positive and increasing $rS'(r)$ and for given positive numbers γ and δ

$$\frac{1}{r}\frac{d}{dr}(rS'(r)) \leq r^{\gamma}(S(r))^{1+\delta} \qquad \|,$$

where $\|$ means that the inequality holds for all r outside an open set I depending on γ, δ such that $\int_I \frac{dt}{t} < \infty$. When the Calculus Lemma is used to get an inequality, the inequality holds only with the condition designated by the notation $\|$. However, for notational simplicity we suppress the notation $\|$ in the rest of this paper.

From this Calculus Lemma and the concavity of the logarithmic function we obtain

$$\mathcal{A}_r(\log\frac{1}{|f^*s|^2(\log|f^*s|^2)^2}) \leq \gamma \log r + (1+\delta)\log\mathcal{I}_r(\frac{1}{|f^*s|^2(\log|f^*s|^2)^2}).$$

We would have gotten $\delta_f(Zs) = 0$ if the numerator $|f^*Ds|^2$ were missing from $\mathcal{I}_r(2\frac{|f^*Ds|^2}{|f^*s|^2(\log|f^*s|^2)^2})$. The contribution to the defect in value distribution theory comes from the existence of this numerator. In the case of a compact Riemann surface when one assumes that the divisor Zs of s has no multiple points,

$$|f^*Ds|^2 = |f_z|^2(f^*|Ds|^2) \geq (c_1 - c_2|f^*s|^2)\ |f_z|^2$$

where c_1 and c_2 are positive constants and f_z is the differential of f with respect to the coordinate z of $\mathbb{C}$. We now use

$$\mathcal{A}_r(\log|f_z|^2) - \log|f_z|^2(0) = \mathcal{I}_r(Zf_z) + \mathcal{I}_r(\theta_{K_M})$$

which comes from integrating twice $\frac{i}{2\pi}\partial\bar{\partial}\log|f_z|^2$, where Zf_z is the zero-set of f_z regarded as a (1,1)-current on $\mathbb{C}$.

$$\mathcal{A}_r\left(\log\frac{1}{|f^*s|^2(\log|f^*s|^2)^2}\right)$$

$$= \mathcal{A}_r\left(\log\frac{|f_z|^2}{|f^*s|^2(\log|f^*s|^2)^2}\right) - \log|f_z|^2(0) - \mathcal{I}_r(Zf_z) - \mathcal{I}_r(\theta_{K_M})$$

$$\leq \gamma \log r + (1+\delta)\log\mathcal{I}_r\left(\frac{|f_z|^2}{|f^*s|^2(\log|f^*s|^2)^2}\right)$$

$$- \log|f_z|^2(0) - \mathcal{I}_r(Zf_z) - \mathcal{I}_r(\theta_{K_M})$$

We now use the inequality $|f^*Ds|^2 \geq (c_1 - c_2|f^*s|^2)\ |f_z|^2$ to get

$$\mathcal{A}_r\left(\log\frac{1}{|f^*s|^2(\log|f^*s|^2)^2}\right)$$

$$\leq C_1 + \varepsilon\mathcal{I}_r(|f_z|^2) - \mathcal{I}_r(Zf_z) - \mathcal{I}_r(\theta_{K_M}) +$$

$$C_2\log r + (1+\delta)\log\mathcal{I}_r\left(\frac{|f^*Ds|^2}{|f^*s|^2(\log|f^*s|^2)^2}\right),$$

where C_1, C_2 are positive numbers, from which we obtain the defect relation $\delta_f(Zs) \leq \sigma$, where σ is any positive number such that $\sigma\,\theta_L \geq \theta_{K_M^{-1}}$ everywhere on M. This is the defect relation of Nevanlinna. The key point is that we have the inequality $|f^*Ds|^2 \geq (c_1 - c_2|f^*s|^2)\ |f_z|^2$ to take care of the troublesome denominator in the boundary integral. When the complex dimension of the target manifold M is $n > 1$, f^*Ds is the inner product of the vector Ds and the vector f_z and we cannot have such an inequality. However, when f is a holomorphic map from $\mathbb{C}^n$ to M with both the domain and target manifolds of equal dimension, one can use the domination of the geometric mean by the arithmetic mean to handle the difficulty in the following way to get the defect relation of Carlson and Griffiths [4]. Let η and ω be respectively the Kähler forms of $\mathbb{C}^n$ and M and we interprete B_r as the open ball of radius r in $\mathbb{C}^n$ centered at the origin and $\mathcal{A}_r(g)$ as $\frac{1}{4\pi r^{2n-1}}\int_{\partial B_r} g$ and $\mathcal{I}_r(\alpha)$ as

$$\mathcal{I}_r(\alpha) = \int_{\rho=0}^{r}\frac{d\rho}{\rho^{2n-1}}\int_{B_\rho}\alpha \wedge \frac{\eta^{n-1}}{(n-1)!}\ .$$

When g is a function, $\mathfrak{I}_r(g)$ means $\mathfrak{I}_r(g\,\eta)$. As before we have

$$\begin{aligned}&\mathfrak{A}_r(\log\frac{1}{|f^*s|^2(\log|f|^2)^2}) - C\\ &\le (1-\varepsilon)\mathfrak{I}_r(f^*\theta_L) - (1+\varepsilon)\mathfrak{I}_r(Zf^*s) - \varepsilon\mathfrak{I}_r(f^*\omega) +\\ &\quad \mathfrak{I}_r(2\frac{|f^*Ds|^2}{|f^*s|^2(\log|f^*s|^2)^2} + \varepsilon\, f^*\omega).\end{aligned}$$

From the domination of the geometric mean by the arithmetic mean of the eigenvalues of $2\frac{|f^*Ds|^2}{|f^*s|^2(\log|f^*s|^2)^2} + \varepsilon\, f^*\omega$ with respect to η, one concludes that for some positive constant c depending on ε

$$\mathfrak{I}_r(2\frac{|f^*Ds|^2}{|f^*s|^2(\log|f^*s|^2)^2} + \varepsilon\, f^*\omega) \ge c\ \mathfrak{I}_r\left[(\frac{|J_f|^2}{|f^*s|^2(\log|f^*s|^2)^2})^{\frac{1}{n}}\right],$$

where J_f is the Jacobian determinant of f. The same conclusion holds when $|f^*s|$ is replaced by $\Pi_{j=1}^q|f^*s_j|^2$ where each s_j is a holomorphic section of L and the zero-sets of s_j $(1 \le j \le q)$ are nonsingular and intersect normally. Thus from the concavity of logarithm and the Calculus Lemma one has the defect relation of Carlson and Griffiths that $\Sigma_{j=1}^q \delta_f(Zs_j)$ does not exceed any positive number σ that satisfies $\sigma\,\theta_L \ge \theta_{K_M^{-1}}$ everywhere on M.

In our case of unequal dimensions for the domain and target manifolds we need a new method to handle the estimate of the boundary term. We have a holomorphic map f from $\mathbb{C}$ to a compact complex manifold M of complex dimension two. We sketch the main steps of this method, first using a rather inaccurate, greatly simplified, intuitive description and then adding modifications to make the description accurate. We use the notations of Theorem 1 with $k = 1$ and the subscript j is dropped.

Step 1. Use the Poincaré-Lelong formula for $\frac{i}{2\pi}\partial\bar\partial\log\frac{1}{|f^*s|^2(\log|f^*s|^2)^2}$ to get an estimate of $\mathfrak{I}_r(\frac{|f^*Ds|^2}{|f^*s|^2(\log|f^*s|^2)^2})$. Here and also for the rest of the sketch of the proof, by an estimate of an expression we mean that the expression is of the order $o(\mathfrak{I}_r(f^*\theta_L))$.

Step 2. Use the Poincaré-Lelong formula for $\frac{i}{2\pi}\partial\bar\partial\log\frac{1}{|f^*Ds|^2(\log|f^*Ds|^2)^2}$ to get an estimate of $\mathfrak{I}_r(\frac{|D(f^*Ds)|^2}{|f^*Ds|^2})$.

Step 3. In terms of local coordinates we have $f^*Ds = f_z^\alpha D_\alpha s$ and $D(f^*Ds) = (D_z f_z^\alpha)D_\alpha s + f_z^\alpha f_z^\beta D_\alpha D_\beta s$. (Here and in the rest of this paper the summation convention of summing over repeated indices is used.) When the "second fundamental form" of Zs is zero with respect to the connection (that is, $D_\alpha D_\beta s$ is a linear combination of $D_\alpha s$, $D_\beta s$, and s with smooth coefficients), we can use Cramer's rule to solve for $D_1 s$, $D_2 s$ in terms of f^*Ds and $D(f^*Ds)$ and $f_z \wedge D_z f_z$ (which is given by the determinant of the matrix with the two row vectors f_z and $D_z f_z$). When Zs is nonsingular, by using the estimates of Steps 1 and 2 and the result just obtained from Cramer's rule and Hölder's inequality, we get an estimate of $\mathfrak{I}_r\left[\left(\frac{|f_z \wedge D_z f_z|^2}{|f^*s|^2(\log|f^*s|^2)^2}\right)^\mu\right]$ for some positive number μ.

Step 4. Use the Poincaré-Lelong formula for $\frac{i}{2\pi}\partial\bar\partial\log|f_z \wedge D_z f_z|^2$ together with the concavity of the logarithm and the Calculus lemma we get, just in the case of the Nevanlinna defect relation, an estimate of the troublesome boundary term and thereby a defect relation.

Now we would like to point out the inaccuracies in the sketch of the above steps and give the modifications needed to remove such inaccuracies. First, when the Poincaré-Lelong formula is used in Step 1, to get useful results one needs to have a *holomorphic* f^*Ds. We would like to use a *holomorphic* connection of L for this purpose instead of the complex metric connection of L from the Hermitian metric of L. Since a holomorphic connection for L in general does not exists, we have to settle for a meromorphic one. Choose a positive Hermitian holomorphic line bundle F_1 over M such that for every nonzero holomorphic section t_1 there exists a meromorphic connection for L whose product with t_1 is holomorphic. This property is satisfied by any sufficiently positive line bundle F_1. We use $\mathfrak{D}$ to denote covariant differenitiation with respect to this connection of L. In Step 1 we have to substitute $f^*(t_1\mathfrak{D}s)$ for f^*Ds. Another inaccuracy is that whereas in Step 1 we can rescale the Hermitian metric of L to guarantee that $|s|$ is sufficiently small on M so that $\frac{2}{\log|f^*s|^2} \le \varepsilon$ on $\mathbb{C}$, we cannot in Step 2 make sure that $\frac{2}{\log|f^*(t_1\mathfrak{D}s)|^2} \le \varepsilon$ on $\mathbb{C}$. Thus we have to use

$$\frac{i}{2\pi}\partial\bar{\partial}\log\frac{1}{|f^*(t_1\mathfrak{D}s)|^2(A + (\log|f^*(t_1\mathfrak{D}s)|^2)^2}$$

in the Poincare-Lelong formula of Step 2 with some positive number A. Again in Step 2 we need a holomorphic (or at least a meromorphic connection for $F_1 \otimes L$). Choose a positive Hermitian holomorphic line bundle F_2 over M such that for every nonzero holomorphic section t_2 there exists a meromorphic connection for $F_1 \otimes L$ whose product with t_2 is holomorphic. Again this property is satisfied by any sufficiently positive line bundle F_2. We again use $\mathfrak{D}$ to denote covariant differenitiation with respect to this connection of $F_1 \otimes L$. The use of the modified Poincaré-Lelong formula in Step 2 gives an estimate of $\mathfrak{T}_r(\frac{|f^*(t_2\mathfrak{D}(t_1\mathfrak{D}s))|^2}{|f^*(t\mathfrak{D}s)|^2(A + (\log|f^*(t\mathfrak{D}s)|^2)^2})$. From this we get, by simple algebraic inequalities and the earlier estimates, an estimate of $\mathfrak{T}_r\left[(\frac{(|f^*(t_2\mathfrak{D}(t_1\mathfrak{D}s))|^2)^\lambda}{|f^*s|^2(\log|f^*s|^2)^2})^\mu\right]$ for suitable $\lambda > 1$ and $\mu > 0$. The presence of $\lambda > 1$ is needed because the factor $A + (\log|f^*(t_1\mathfrak{D}s)|^2)$ requires compensation of higher vanishing order of $f^*(t_2\mathfrak{D}(t_1\mathfrak{D}s))$. Moreover, we can get the estimate only for sufficiently small positive μ, because transforming a product into a sum in an inequality necessitates using a sum of powers of the factors in the product. In Step 4 when the Poincaré-Lelong formula is used for $\frac{i}{2\pi}\partial\bar{\partial}\log|f_z \wedge D_z f_z|^2$, we need the *holomorphicity* of $f_z \wedge D_z f_z$ which in general is of course not holomorphic. So we have to introduce a holomorphic connection for the tangent bundle T_M of M. Again since holomorphic connections for T_M in general do not exist, we have to settle for a meromorphic one. We have to assume that there exists a meromorphic connection $\Gamma^{\gamma}_{\alpha\beta}$ for T_M and a holomorphic Hermitian line bundle F over M with a non-identically-zero global holomorphic section t so that $t\Gamma^{\gamma}_{\alpha\beta}$ is holomorphic on M. We denote also by $\mathfrak{D}$ the covariant differentiation in this case. Because in Step 3 we need the vanishing of the "second fundamental form" of Zs with respect to the connection of T_M, in addition we have to assume that locally on M for any fixed α, β, the component $t\mathfrak{D}_\alpha \partial_\beta s$ is a linear combination of $\partial_\alpha s$, $\partial_\beta s$, and s with smooth coefficients. The vanishing of the "second fundamental form" of Zs together with the estimate of $\mathfrak{I}_r\left[\left(\frac{(|f^*(t_2\mathfrak{D}(t_1\mathfrak{D}s))|^2)^\lambda}{|f^*s|^2(\log|f^*s|^2)^2}\right)^\mu\right]$ gives us an estimate of $\mathfrak{I}_r\left[\left(\frac{(|t_1 t_2(t\mathfrak{D}_z f^\alpha_z)(t_1\mathfrak{D}_\alpha s)|)^\lambda}{|f^*s|^2(\log|f^*s|^2)^2}\right)^\mu\right]$ for suitable $\lambda > 1$ and $\mu > 0$. The use of Cramer's rule gives us an estimate of $\mathfrak{I}_r\left[\left(\frac{|f_z \wedge t\mathfrak{D}_z f_z|^2}{|f^*s|^2(\log|f^*s|^2)^2}\right)^\mu\right]$ for suitable $\mu > 0$. The use of the Poincare-Lelong formula for $\frac{i}{2\pi}\partial\bar\partial\log|f_z \wedge t\mathfrak{D}_z f_z|^2$ yields the defect relation $\delta_f(Zs) \le$ any positive number σ that satisfies $\sigma\,\theta_L \ge \theta_{FK_M^{-1}}$. The line bundle F enters in the condition for σ because of the presence of t in $\frac{i}{2\pi}\partial\bar\partial\log|f_z \wedge t\mathfrak{D}_z f_z|^2$.

§4. Formulation of Higher Dimensional Case.

In the case of a holomorphic map f from $\mathbb{C}$ to a compact complex manifold M of complex dimension $n \ge 2$, the same method of proof can be used to get the analog of Theorems 1 and 2. As in Theorems 1 and 2 we assume that we have L, $\Gamma^{\gamma}_{\alpha\beta}$, F, t, s_j, k satisfying the same conditions. Again assume that $\lim_{r\to\infty} \frac{\log r}{\mathfrak{I}_r(f^*\theta_L)} = 0$. Assume the nondegeneracy condition that the map

$$f_z \wedge t\mathfrak{D}_z f_z \wedge t\mathfrak{D}_z(t\mathfrak{D}_z f_z) \wedge \cdots \wedge (t\mathfrak{D}_z)^{n-1} f_z$$

is not identically zero. Then the sum $\Sigma_{j\in J}\,\delta_f(Zs_j)$ of the defects of

f for the divisors Zs_j $(j \in J)$ does not exceed any nonnegative number $k\sigma$ such that $\sigma\theta_L \geq \theta_{K_M^{-1}F^{n(n-1)/2}}$.

For the case of a holomorphic map f from $\mathbb{C}^m$ to a compact complex manifold M of complex dimension $n \geq m$ we have similar results. The nondegeneracy condition is that

$$\bigwedge_{\substack{1\leq i_1\leq\cdots\leq i_\ell\leq m\\ 0\leq\ell\leq n-m}} t\mathfrak{D}_{z_{j_1}}(t\mathfrak{D}_{z_{j_2}})\cdots(t\mathfrak{D}_{z_{j_\ell}})(f_{z_1}\wedge\cdots\wedge f_{z_m})$$

as a map from $\mathbb{C}^m$ to the line bundle $\Lambda^{\binom{n}{m}}(\Lambda^m T_M)$ is not identically zero, where $z_1,\cdots,z_m$ are the coordinates of $\mathbb{C}^m$ and the subscript z_j means differentiation with respect to z_j. Let

$$N = \sum_{\ell=0}^{n-m} \ell \binom{n+\ell-1}{\ell}.$$

Then the sum $\Sigma_{j\in J}\,\delta_f(Zs_j)$ of the defects of f for the divisors Zs_j $(j \in J)$ does not exceed any nonnegative number $k\sigma$ such that $\sigma\theta_L \geq \theta_{(\Lambda^{\binom{n}{m}}(\Lambda^m T_M))\otimes F^N}$.

REFERENCES.

1. L. Ahlfors, The theory of meromorphic curves, Acta Soc. Sci. Fenn, Nova Ser. A 3(4) (1941), 171-183.
2. A. Biancofiore, A hypersurface defect relation for a class of meromorphic maps, Trans. Amer. Math. Soc. 270 (1982), 47-60.
3. E. Bishop, Conditions for analyticity of certain sets, Michigan Math. J. 11 (1964), 289-304.
4. J. Carlson and Ph. Griffiths, Defect relation for equidimensional holomorphic mappings between algebraic varieties, Ann. of Math. 95 (1972), 557-584.
5. M. Cowen and Ph. Griffiths, Holomorphic curves and metrics of nonnegative curvature, J. Analyse Math. 29 (1976), 93-153.
6. E. El Mir, Sur le prologement des courants positifs fermés de masse finie, C.R. Acad. Sci. Paris 264 (1982), 181-184.

7. M. Green, Some Picard theorems for holomorphic maps to algebraic varieties, Amer. J. Math. 97 (1975), 43-75.

8. J. Noguchi, Holomorphic curves in algebraic varieties, Hiroshima Math. J. 7 (1977), 833-853.

9. R. Remmert and K. Stein, Über die wesentlichen Singularitäten analytischer Mengen, Math. Ann. 126 (1953), 263-306.

10. B. Shiffman, Nevanlinna defect relations for singular divisors, Invent. Math. 31 (1975), 155-182.

11. B. Shiffman, On holomorphic curves and meromorphic maps in projective spaces, Indiana Univ. Math. J. 28 (1979), 627-641.

12. N. Sibony, Quelques problèmes de prolongement de courants en analyse complexe, Duke Math. J. 52 (1985), 157-197.

13. H. Skoda, Prolongement des courants, positifs, fermés, de masse finie, Invent. Math. 66 (1982), 361-376.

14. W. Stoll, The growth of the area of a transcendental analytic set I, II, Math. Ann. 156 (1964), 144-170.

15. W. Stoll, Value distribution of holomorphic maps into compact, complex manifolds, Lecture Notes in Mathematics 135 (1970) Springer-Verlag.

16. W. Stoll, The Ahlfors-Weyl theory of meromorphic maps on parabolic manifolds, Lecture Notes in Mathematics 981 (1983), 101-219.

17. P. Thullen, Über die wesentlichen Singularitäten analytischer Funktionen und Flächen im Raume von n komplexen Veränderlichen, Math. Ann. 111 (1935), 137-157.

18. H. Weyl and J. Weyl, Meromorphic curves, Ann. of Math. 39 (1938), 516-538.

Author's address: Department of Mathematics, Harvard University, Cambridge, Massachusetts 02138, U.S.A.

On the Adjunction Theoretic Structure of Projective Varieties

Andrew John Sommese
Department of Mathematics
University of Notre Dame
Notre Dame, Indiana, 46556 U.S.A.

To Rebecca

Let X be a normal n dimensional projective variety. Let L be an ample line bundle on X, i.e. assume that there is a holomorphic finite to one mapping

$$\varphi : X \longrightarrow \mathbb{P}_{\mathbb{C}}$$

such that $L^t \cong \varphi^* \mathcal{O}_{\mathbb{P}}(1)$ for some positive integer t. Let K_X denote the dualizing sheaf on X and let

$$K_X{}^N = (\underbrace{K_X \otimes \cdots \otimes K_X}_{N \text{ copies}})^{**}.$$

We are interested in studying the pair (X,L) in terms of $K_X + \tau L$ where τ is a real number.

Given a pair (X,L) as above we define its **spectral value**, $\sigma(X,L)$, as the smallest real number τ such that given any fraction $p/q > \tau$,

$$\Gamma(K_X{}^N \otimes L^{(n+1-p/q)N}) = 0 \text{ for all integers } N > 0 \text{ with } q \mid N.$$

The smallest spectral value is 0.

In this paper we consider pairs (X,L) that satisfy the following conditions:

a) X is $\mathbb{Q}$-Gorenstein, i.e. there is a positive integer $r > 0$ called the index of X such that $K_X{}^r$ is invertible,

b) the set of non-rational singularities of X is finite,

c) L is spanned by global sections, i.e. the integer N in the definition of ample above can be taken to be 1,

and

d) K_X is invertible off of a finite set.

The first two assumptions are quite natural; examples, e.g. (0.2.4) show that our results fail without them. The third assumption is mild since we are mainly interested in applying these results to subvarieties of projective spaces. The last assumption seems unnecessary but we haven't been able to remove it.

The following table organizes our results for $\sigma(X,L) \leqslant 2$.

<table>
<tr><th>σ(X,L)</th><th>Structure</th><th>Index</th><th>Singularities</th></tr>
<tr><td>0</td><td>$(\mathbb{P}^n, \mathcal{O}_{\mathbb{P}}(1))$</td><td>1</td><td>smooth</td></tr>
<tr><td rowspan="2">1</td><td>quadric</td><td>1</td><td>local complete intersection</td></tr>
<tr><td>scroll over a smooth curve</td><td>1</td><td>smooth</td></tr>
<tr><td>$2 - 2/e$ for $e \geqslant 3$</td><td>non-degenerate cone over $(\mathbb{P}^1, \mathcal{O}_{\mathbb{P}}(e))$ for $e \geqslant 3$</td><td>e</td><td rowspan="2">quotient singularities
in particular rational</td></tr>
<tr><td>3/2</td><td>possibly degenerate generalized cones over $(\mathbb{P}^2, \mathcal{O}_{\mathbb{P}}(2))$</td><td>1 if $n = 2$
2 if $n \geqslant 3$</td></tr>
<tr><td rowspan="3">2</td><td>Del Pezzo Variety</td><td>1</td><td>Gorenstein</td></tr>
<tr><td>quadric bundle over a smooth curve</td><td>1</td><td>local complete intersection</td></tr>
<tr><td>scroll over a surface $n \geqslant 3$</td><td></td><td>Gorenstein rational singularities</td></tr>
<tr><td>> 2</td><td>For all remaining varieties $K_X \otimes L^{n-1}$ is semi-ample and big</td><td></td><td></td></tr>
</table>

All of the (X,L) with $\sigma(X,L) < 2$ can be characterized in a number of ways. The following statements are equivalent for $n \geqslant 2$:

a) $\sigma(X,L) < 2$,

b) $h^1(\mathcal{O}_X) = g(L)$ where $2g(L)-2 = (K_X+(n-1)L)\cdot \underbrace{L \cdot \cdots \cdot L}_{n-1 \text{ times}}$,

c) $h^{n-1}(K_X) = g(L)$,

d) $h^0(K_X \otimes L^{n-1}) = 0$.

The two key new technical ingredients in the proofs of these results are :

1) F. Sakai's paper [Sa3] which very thoroughly studied the properties of pairs (X,L) where L is an ample line bundle on a normal surface. These results played a major role in many of our inductive proofs,

2) A somewhat surprising result that numerical effectivity ascends off an ample divisor under weak conditions,

(2.1) **Theorem**. _Let X be a normal projective $\mathbb{Q}$-Gorenstein variety with both_ Irr(X) _and the set of non-Gorenstein points finite. Let L be an ample line bundle on X and let_ $A \in |L|$. _Assume that X is Gorenstein in a neighborhood of A. Let τ be a real number_ $\geqslant 2$. _If_ $(K_X + \tau L)_A$ _is numerically effective then_ $K_X + \tau L$ _is numerically effective._

To analyze the pairs (X,L) with $K_X \otimes L^{n-1}$ semi-ample and big we are forced to impose extra conditions. For simplicity we assume that :

α) X is Gorenstein, i.e. K_X is invertible and X is Cohen-Macaulay,

β) cod Sing(X) $\geqslant 3$, and the locus Irr(X) of non-rational singularities is both finite and of codimension $\geqslant 4$.

We define a **reduction (X',L') of a pair (X,L)** as a pair (X',L') where L' is an ample line bundle on a normal projective variety, X', such that :

a) there is a map $\pi : X' \longrightarrow X$ expressing X' as X with a finite set $F \subset \mathrm{reg}(X)$ blown up,

b) $L \cong \pi^* L' \otimes [\pi^{-1}(F)]^{-1}$ or equivalently $K_X \otimes L^{n-1} \cong \pi^*(K_{X'} \otimes L'^{n-1})$.

The following table summarizes our results for $\sigma(X,L) \leqslant 3$.

<table>
<tr><th>$\sigma(X,L)$</th><th colspan="2">Structure of (X',L')</th></tr>
<tr><td rowspan="3">5/2</td><td colspan="2">There is a holomorphic surjection $\varphi : X' \longrightarrow C$ onto C, a smooth curve where the general fibre of φ is $(\mathbb{P}^2, \mathcal{O}_{\mathbb{P}^2}(2))$ and $K_{X'}^2 \otimes L'^3 \cong \varphi^* \mathcal{L}$ for an ample line bundle $\mathcal{L}$ on C.</td></tr>
<tr><td colspan="2">$(\mathbb{Q}, \mathcal{O}_{\mathbb{Q}}(2))$ where $(\mathbb{Q}, \mathcal{O}_{\mathbb{Q}}(1))$ is a 3 dimensional quadric</td></tr>
<tr><td colspan="2">$(\mathbb{P}^4, \mathcal{O}_{\mathbb{P}^4}((2))$</td></tr>
<tr><td>8/3</td><td colspan="2">$(\mathbb{P}^3, \mathcal{O}_{\mathbb{P}^3}(3))$</td></tr>
<tr><td rowspan="4">3</td><td>$K_{X'} \otimes L'^{n-2} \cong \mathcal{O}_{X'}$</td><td>Characterizations</td></tr>
<tr><td>Del Pezzo bundle over a smooth curve</td><td rowspan="4">a) $K_{X'} \otimes L'^{n-2}$ is nef
b) $K_{X'} \otimes L'^{n-2}$ is semi-ample
c) $h^0((K_{X'} \otimes L'^{n-2})^N) > 0$ for some integer $N > 0$</td></tr>
<tr><td>Quadric bundle over a surface</td></tr>
<tr><td>Scroll over a 3 fold - here $n \geq 4$</td></tr>
<tr><td>> 3</td><td>For all remaining pairs $K_{X'} \otimes L'^{n-2}$ is semi-ample and big</td></tr>
</table>

The key techniques needed to show the above results are come from :

1) Fania and Sommese [(F+S)1], Lipman and Sommese [L+S], and Sommese [So6],

2) the basepoint free theorem of Kawamata [Ka3].

It should be noted that the hypothesis of spannedness of L can be relaxed to the hypothesis that there are dim X-2 sections of L that have a normal surface as there scheme theoretic intersection.

The scheme of the paper is as follows.

In § 0, I collect background material and prove 'folklore' facts for which I know no good reference.

In § 1 I define spectral values and spectral classes, and prove a few general facts about the concepts.

In § 2 I prove the key results on ascent of nefness and semi-ampleness.

In § 3 I define some special classes of varieties. I also collect and prove some facts about them.

In § 4 I prove the structure theorems for the spectral class $\mathcal{K}_2$.

In § 5 I prove the structure theorems for the spectral class $\mathcal{K}_3$.

The use of adjunction to study the relation between a variety and properties stemming from its projective embedding is very classical (see [Ro] and [So2]). Much of the modern work on questions of the sort considered in this paper has it start in [So2] where the spannedness of $K_X \otimes L$ for a very ample line bundle L on a smooth surface was studied (see also [Sa1] for a parallel line of research that influenced [So2]).

After [So2] various aspects of the theory of $K_X \otimes L^{n-1}$ for L ample and spanned on a smooth n fold and of $K_X \otimes [A]$ for a smooth ample divisor on a smooth threefold were worked out in [So3] to [So7]. These papers culminated in [So6] which was never published because N. Shepherd-Barron had shown me how to prove much stronger results by using the then new theory of Mori [M] (see also [B+P], [Ka2], [Ka2]).

In [L+S] and [(F+S)1] the theory of [So6] was redone for Gorenstein threefolds with isolated singularities.

In [L+P] it was shown that the results of [So2] regarding asymptotic properties of $(K_X \otimes L)^N$ for $N >> 0$ held for ample line bundles L on X, a smooth surface. In [So8], I showed how to prove the same results with the smoothness of X relaxed to the property of X being normal and Gorenstein. I also made a study of the spannedness properties of $K_X \otimes L$ on Gorenstein surfaces. This paper was followed by [So9] where the analogous properties of $K_X \otimes L^{n-1}$ were studied on Gorenstein n folds with an ample and spanned line bundle L.

In [Sa3] the results of [L+P], [So6], and [So8] on the asymptotic properties of $(K_X \otimes L)^N$ for $N >> 0$ were generalized for normal surfaces by F. Sakai as an application of his general theory [Sa2]. For $\mathbb{Q}$-Gorenstein surfaces these results were as strong as the results of [L+P] or [So8] except that a new series of pairs (X,L) with $h^0((K_X \otimes L)^N) = 0$ for all $N > 0$ turned up.

Inspired by Sakai's work [Sa3], I discovered the key result (2.1) from which this paper grew. As I was writing the original version of this paper I received the research notice [F6] from T. Fujita and a letter from M. Beltrametti announcing very similar results to mine for pairs (X,L) such that L is ample on X, a smooth manifold; Beltrametti pointed out that P. Ionescu has also obtained similar results. After this paper was typed

I received the preprints "Generalized adjunction and applications" from Ionescu and "On polarized manifolds whose adjoint bundles are not nef" from Fujita. Though their line bundles are more general than mine, their X must for the most part be smooth. This is because they depend heavily on forms of Mori theory. On the other hand I mainly use various modifications of the classical Fano-Morin adjunction process [Ro] and induction from Sakai's results.

The methods of this paper still yield strong (though only birational) results when applied to normal varieties minus their singular sets; this will be carried out elsewhere.

I would like to thank F. Sakai for inspiring me by his preprint [Sa3] and for suggesting improvements of both the results and style of this paper. I would like to thank the National Science Foundation (DMS 8200 629 and DMS 8420 315), the University of Notre Dame, and the Max Planck Institut für Mathematik in Bonn, West Germany for their support.

I would like to thank the organizers of the Göttingen conference for inviting me to it and I would like to express the appreciation that many of the participants felt for a conference with a schedule designed for human mathematicians.

§ 0 Notation and Background Material

(0.0) We work over the complex numbers. All spaces are complex analytic and all maps are holomorphic. By **variety** we mean an irreducible and reduced complex analytic space. If X is a complex analytic space, we denote its holomorphic structure sheaf by $\mathcal{O}_X$. We do not distinguish notationally between a locally free coherent analytic sheaf and its associated holomorphic vector bundle.

If X is a connected complex manifold then we have the dualizing sheaf $K_X = \wedge^n T_X^*$ where $n = \dim X$ and T_X^* is the cotangent bundle of X. If X is a normal variety then the dualizing sheaf $K_X = j_* K_{Reg(X)}$ where $j : Reg(X) \longrightarrow X$ is the inclusion of the smooth points of X into X. In this case K_X is a reflexive rank one sheaf on X. When X is normal and projective and $\tau \in \mathbb{R}$ we use τK_X to denote the Weil divisor obtained by multiplying τ times the Weil divisor K_X. I have found [Re] and [A+K] helpful references for these matters.

Given an effective normal Cartier divisor A on a normal Cohen-Macaulay variety X we have by ([A+K], pg. 7):

$$(K_X \otimes [A]) \cong K_A.$$

(0.0.1) Let X be a variety and let $p : \bar{X} \longrightarrow X$ be a resolution of singularities, i.e. $\bar{X}$ is a complex manifold and p is a surjective holomorphic map which gives a biholomorphism from $\bar{X} - p^{-1}(\mathrm{Sing}(X))$ to $X - \mathrm{Sing}(X)$. The Leray sheaves $p_{(i)}(\mathcal{O}_{\bar{X}})$ for $i \geq 0$ are independent of the resolution; we denote them by $\mathbf{\mathcal{S}_i(X)}$. X is normal if and only if $\mathcal{S}_0(X) \cong \mathcal{O}_X$.

Assume that X is normal. If $\mathcal{S}_i(X) = 0$ for $i > 0$ then the singularities of X are said to be rational. It is a theorem of Kempf ([Ke], pg. 50) that X has rational singularities if and only if X is Cohen-Macaulay and $p_*(K_{\bar{X}}) \cong K_X$. We denote by **Irr(X)**, the **irrational locus of X**, which is the union of the supports of the sheaves $\mathcal{S}_i(X)$ for $i > 0$.

Given a reflexive rank one sheaf $\mathcal{S}$ on X and an integer $t > 0$, we use $\mathcal{S}^t$ to denote

$$\underbrace{(\mathcal{S} \otimes \cdots \otimes \mathcal{S})}_{-\,t\text{ times}\,-}{}^{**}$$

and $\mathcal{S}^{-t}$ to denote $\mathcal{S}^{*\,t}$. It is convenient to set $\mathcal{S}^0 \cong \mathcal{O}_X$.

(0.1) If $\mathcal{S}^t$ is invertible for some $t > 0$ and X is a normal projective variety then $c_1(\mathcal{S}) \in H^2(X,\mathbb{Q})$ is well defined as $c_1(\mathcal{S}^t)/t$. Such an $\mathcal{S}$ is said to be **big** if $c_1(\mathcal{S})^{\dim X} > 0$ and to be **numerically effective** (or **nef** for short) if $c_1(\mathcal{S})[C] \geq 0$ for all effective curves C on X. If there is a $t > 0$ such that $\mathcal{S}^t$ is invertible and spanned by global sections then $\mathcal{S}$ is said to be **semi-ample**. By going to a large enough positive multiple N of t we can then assume that there is a holomorphic map $\varphi : X \longrightarrow \mathbb{P}_{\mathbb{C}}$ with connected fibres and normal image $Y = \varphi(X)$ such that $\mathcal{S}^N = \varphi^* \mathcal{O}_{\mathbb{P}}(1)$.

If $K_X{}^t$ is invertible for some integer $t > 0$ then X is said to be **$\mathbb{Q}$-Gorenstein** and the the smallest such t is called the **index** of X; in this case given any line bundle L on X and any normal $A \in |L|$ and any positive integer multiple N of t

$$(K_X{}^N \otimes L^N)_A \cong K_A{}^N.$$

If the index is one and X is Cohen-Macaulay then X is said to be **Gorenstein**.

(0.2) We need the following vanishing theorem ([(F+L)2]; theorem (0.2)) due to Picard, Kodaira, Ramanujam, Mumford, Grauert and Riemenschneider [G+R], Kawamata [Ka1] and Viehweg [V], and Kempf [Ke]. See [Sh+So] for a discussion of similar results.

(0.2.1) **Theorem**. Let L be a nef and big line bundle on a normal projective variety X. Then :

a) $H^i(X, L^{-1}) = 0$ _for_ $i < \min\{\dim X, 2\}$,
b) $H^i(X, K_X \otimes L) = 0$ _for_ $i > \max\{0, \dim \mathrm{Irr}(X)\}$.

Proof. a) is a simple variant of Mumford's vanishing theorem that uses [Ka1] and [V] in place of the usual Kodaira vanishing theorem (cf. [Sh+So]).

To see b), let $p : \bar{X} \longrightarrow X$ be a projective desingularization of X. Consider the exact sequence :

(0.2.1*) $0 \longrightarrow p_*(K_{\bar{X}}) \otimes L \longrightarrow K_X \otimes L \longrightarrow \mathcal{S} \otimes L \longrightarrow 0.$

By Kempf's theorem support$(\mathcal{S} \otimes L) \subset \mathrm{Irr}(X)$. Therefore by the long exact cohomology sequence associated to *), the theorem will follow from $H^i(p_*K_{\bar{X}} \otimes L) = 0$ for $i > \max\{0, \dim \mathrm{Irr}(X)\}$. Since :

$$P_{(i)}(K_{\bar{X}} \otimes p^*L) \cong P_{(i)}(K_{\bar{X}}) \otimes L = 0 \text{ for } i > 0$$

by the Grauert- Riemenschneider vanishing theorem, theorem (0.2.1b) will follow from the Leray spectral sequence for p and p^*L, and the fact that $H^i(K_{\bar{X}} \otimes p^*L) = 0$ for $i > 0$. This last fact follows immediately from the Kawamata-Viehweg vanishing theorem ([Ka1], [V]).

□

(0.2.2) **Corollary**. _Let_ X _be a normal Cohen-Macaulay projective variety with_ Irr(X) _finite. Let_ L _be a nef and big line bundle on_ X. _Then_

$$h^0(K_X \otimes L) \geq \#(\mathrm{Irr}(X))$$

where $\#(\mathrm{Irr}(X))$ _is the set-theoretic number of points in_ Irr(X).

Proof. Consider (0.2.1*) as in the proof of the above theorem. By the same argument as in that theorem we see that $H^1(p_*(K_{\bar{X}} \otimes p^*L)) = 0$ and

$$\Gamma(K_X \otimes L) \longrightarrow \Gamma(\mathcal{S} \otimes L) \longrightarrow 0.$$

Note that $h^0(\mathcal{S} \otimes L) = h^0(\mathcal{S}) \geq \#(\mathrm{Irr}(X))$.

□

(0.2.3) **Corollary**. _Let_ X _be a normal projective variety with_ $\dim X \geq 2$. _Let_ L _be a nef and big line bundle on_ X. _If_ $A \in |L|$ _then_ $h^1(\mathcal{O}_X) \leq h^1(\mathcal{O}_A)$. _Further_ A _is connected (and therefore irreducible if_ A _is also normal). If_ X _is Cohen-Macaulay and_ Irr(X) _is finite then_ _the restriction map_ :

$$r : H^i(\mathcal{O}_X) \longrightarrow H^i(\mathcal{O}_A)$$

is an isomorphism for $i < \dim A$ _and an injection for_ $i = \dim A$.

Proof. Consider :

$$0 \longrightarrow L^{-1} \longrightarrow \mathcal{O}_X \longrightarrow \mathcal{O}_A \longrightarrow 0.$$

By (0.2.1a), $\Gamma(\mathcal{O}_X) \cong \Gamma(\mathcal{O}_A)$ and $h^1 \leqslant h^1(\mathcal{O}_A)$. Since X is irreducible we conclude that $\Gamma(\mathcal{O}_A) \cong \mathbb{C}$ and therefore that A is connected.

If Irr(X) is finite $h^i(K_X \otimes L) = 0$ for $i > 0$ by (0.2.1b). Since X is Cohen-Macaulay we conclude by Serre duality that $h^i(L^{-1}) =$ for $i < \dim X$.

□

The key role that Theorem (0.2.1) plays forces us to assume in our theorems that Irr(X) is finite. This is more than a restriction of technique. The Kodaira vanishing theorem and most of the structure theorems of this paper can fail if Irr(X) is not a finite set - even for normal divisors with "nice" singularities on smooth projective fourfolds. I would like to thank F. Sakai for suggesting the following sort of example (cf. [G+R]).

(0.2.4) **Example**. Let $\mathbf{P} = \mathbb{P}(E) = (\mathbf{E}^* - \mathbb{P}^1)/\mathbb{C}^*$ where

$$\mathbf{E} = \mathcal{O}_{\mathbb{P}^1} \oplus \mathcal{O}_{\mathbb{P}^1} \oplus \mathcal{O}_{\mathbb{P}^1}(1) \oplus \mathcal{O}_{\mathbb{P}^1}(1).$$

Let $\pi : \mathbf{P} \longrightarrow \mathbb{P}^1$ denote the tautological projection, let ξ denote the tautological line bundle such that $\pi_*(\xi) \cong \mathbf{E}$, and let F denote a fibre of π. Let X be a general element of $|d(\xi - F)|$. Note that for $d \geqslant 1$, $\pi_X : X \longrightarrow \mathbb{P}^1$ is a bundle with fibre biholomorphic to a cone in $\mathbb{P}^3$ on a smooth curve of degree d. $K_{\mathbf{P}} = \xi^{-4} \otimes [F]$ and therefore $K_X \cong (\xi^{d-4} \otimes [F]^{1-d})_X$. Note that $L = \xi \otimes [F]$ is very ample. A straightforward calculation shows that for $d \geqslant 3$, $h^1(K_X \otimes L_X) = {}_dC_4$, (where ${}_dC_4$ stands for d choose 4); in particular the Kodaira vanishing theorem fails. Note further that $\xi^{2d-5} \cong K_X \otimes L_X{}^{d-1}$ is spanned by global sections and that the map associated to $\Gamma(\xi_X{}^{2d-5})$ has a three dimensional image with the curve associated to $\mathbf{E} \longrightarrow \mathcal{O}_{\mathbb{P}^1} \longrightarrow 0$ as its only positive dimensional fibre. Therefore the structure theorems of this paper fail rather badly for (X, L_X).

By choosing appropriate branched covers of X, we can construct a similar counterexample which contains no rational curves.

To relax the condition that Irr(X) is finite it seems clear that the ampleness properties of the sheaves $\mathcal{S}_i(X)$ must be better understood.

(0.3) We need a few elementary technical lemmas.

(0.3.1) **Lemma**. *Let* A *be an ample normal Cartier divisor on* X, *a normal projective variety of dimension at least. Let* $\mathcal{L}$ *be a torsion free rank one coherent sheaf on* X *such that* $\mathcal{L}^N$ *is invertible for some* $N > 0$ *and* $\mathcal{L}_A \cong \mathcal{O}_A$. *If* $\mathcal{L}^N \cong \mathcal{O}_X$ *or if* $\dim X \geq 3$ *then* $\mathcal{L} \cong \mathcal{O}_X$.

Proof. Assume that $\mathcal{L}^N \cong \mathcal{O}_X$ and that N is the smallest integer ≥ 1 so that this happens. We now use the cyclic covering trick ([Re], (1.9)). Let $\mathbf{R} = \mathcal{O}_X \oplus \mathcal{L}^{-1} \oplus \cdots \oplus \mathcal{L}^{-N}$. $\mathbf{R}$ is a sheaf of rings with the algebra structure given in the obvious way using a fixed trivialization of $\mathcal{L}^N$. We let $\overline{X} = \mathrm{spec}(\mathbf{R})$. It can be checked that $\overline{X}$ is normal and that the natural map $p : \overline{X} \longrightarrow X$ is a finite branched cover that branches only over the set where $\mathcal{L}$ is not invertible. Since $\mathcal{L}_A \cong \mathcal{O}_A$ it follows that p is étale in a neighborhood of $p^{-1}(A)$ and that $p^{-1}(A)$ consists of N disjoint Cartier divisors on $\overline{X}$, each with ample normal bundle. This is absurd if $N > 1$.

We will be done if we show that $\dim X \geq 3$ and $\mathcal{L}_A{}^N \cong \mathcal{O}_A$ imply that $\mathcal{L}^N \cong \mathcal{O}_X$. This follows from Corollary (2.6) of [F2] that restriction gives :

$$0 \longrightarrow \mathrm{Pic}(X) \longrightarrow \mathrm{Pic}(A) \quad \text{if} \quad \dim X \geq 3.$$

□

The next lemma occurs in various degrees of generality in my earlier papers [So1], [So6]; the proof is an easy modification of the proofs of these earlier results.

(0.3.2) **Lemma**. *Let* A *be an effective ample Cartier divisor on a normal variety* X. *Let* $\varphi : X \longrightarrow Y$ *be a holomorphic map onto an analytic set* Y. *If* $\dim \varphi(A) < \dim A$ *then* $\varphi(X) = \varphi(A)$.

Let X be a normal projective variety. Let $p : \overline{X} \longrightarrow X$ be a resolution of singularities of X. We say that the **Albanese mapping is defined** for X if there is a commutative diagram of holomorphic maps :

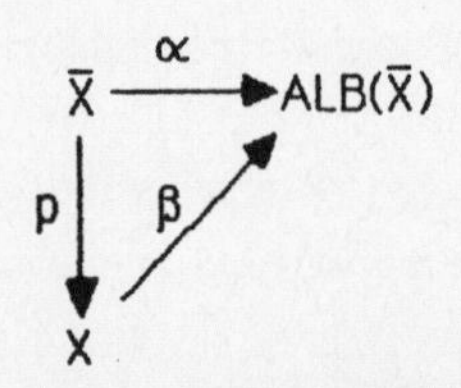

where ALB($\bar{X}$) is the Albanese variety of $\bar{X}$ and α is the Albanese mapping (we suppress basepoints). In this case β and ALB($\bar{X}$) are independent of the resolution. By the **Albanese variety**, ALB(X), we mean ALB($\bar{X}$) with the map β, which we refer to as the **Albanese mapping** of X. The following is a "folklore" lemma.

(0.3.3) **Lemma**. *Let X be a normal projective variety. Then the Albanese mapping of X is defined if* $h^0(\mathcal{S}_1(X)) = 0$, *e.g. if X has rational singularities.*

Proof. Let $p : \bar{X} \longrightarrow X$ be a resolution of singularities and let $\alpha : \bar{X} \longrightarrow$ ALB($\bar{X}$) be the Albanese mapping. It is a consequence of the normality of X that the map α will descend to a holomorphic mapping on X if and only if for each $x \in$ Sing(X) it follows that $\alpha(p^{-1}(x))$ is a point of ALB($\bar{X}$). If this failed then we could pullback an element of $H^1(\mathcal{O}_{ALB(\bar{X})})$ which would restrict to a non-trivial element of $H^1(\mathcal{O}_{p^{-1}(x)})$. In particular we would have a non-trivial element of $\Gamma(\mathcal{S}_1(X))$.

□

In the above argument we only need that the differential from $\Gamma(\mathcal{S}_1(X))$ to $H^2(\mathcal{O}_X)$ in the Leray spectral sequence for p and $\mathcal{O}_{\bar{X}}$ be injective.

(0.4) We need a few Bertini type theorems in somewhat greater generality than I know references for. The first is a generalization of Seidenberg's theorem.

(0.4.1) **Theorem**. *Let X be a normal variety and let L be a line bundle on X spanned by a finite dimensional space* **V** *of global sections. Let* |**V**| *denote the linear space of Cartier divisors associated to* **V**. *There is a Zariski dense set* $U \subset |\mathbf{V}|$ *of divisors D such that* :

a) D *is normal,* Sing(D) ⊂ Sing(X), *and no irreducible component of* Sing(X) *belongs to* D,

b) *for* $i > 0$ *the support of* $\mathcal{S}_i(D)$ *is a subset of the support of* $\mathcal{S}_i(X)$, *and no (possibly embedded) irreducible component of the support of* $\mathcal{S}_i(X)$ *is a subset of* D.

In particular cod Sing(X) ≤ cod Sing(D) *and* cod Irr(X) ≤ cod Irr(D). *If* X *is compact then* U *can be chosen to be Zariski open.*

Proof. The condition that $D \in |\mathbf{V}|$ contains either an irreducible component of Sing(X) or a (possibly embedded) irreducible component of the support of one of the $\mathcal{S}_i(X)$ for some $i > 0$ is an analytic condition, i.e. the condition defines a proper analytic subset of |**V**|. Therefore there is a dense Zariski open set $U' \subset |\mathbf{V}|$ such that if $D \in U'$ then no irreducible (possibly embedded) component of the support of $\mathcal{S}_1(X)$ belongs to D.

Let $p : \bar{X} \longrightarrow X$ be a resolution of X. Since $p^*\mathbf{V}$ spans p^*L we know by the usual Bertini theorem that a dense set of elements of $p^*\mathbf{V}$ are smooth. Let D be one of the dense set, U, of elements of U' such that $\bar{D} = p^{-1}(D)$ is smooth. Since $p : \bar{X} - p^{-1}(\mathrm{Sing}(X)) \longrightarrow X - \mathrm{Sing}(X)$ is a biholomorphism it follows that $\mathrm{Sing}(D) \subset \mathrm{Sing}(X)$ for these D. To finish we must show that D is normal, Let $s \in \mathbf{V}$ be such that $s^{-1}(0) = D$ and let $\bar{s} = s \circ p \in p^*\mathbf{V}$. Consider the sequence :

$$0 \longrightarrow p^*L^{-1} \xrightarrow{\otimes \bar{s}} \mathcal{O}_{\bar{X}} \longrightarrow \mathcal{O}_{\bar{D}} \longrightarrow 0.$$

Consider the long exact sequence of direct image sheaves under p.

$$0 \longrightarrow L^{-1} \longrightarrow \mathcal{O}_X \longrightarrow p_*\mathcal{O}_{\bar{D}} \longrightarrow \mathcal{S}_1(X)\otimes L^{-1} \xrightarrow{\otimes s} \mathcal{S}_1(X) \longrightarrow \mathcal{S}_1(D) \longrightarrow 0.$$

By the choice of D satisfying b) it follows that for all $i > 0$:

$$0 \longrightarrow \mathcal{S}_1(X)\otimes L^{-1} \longrightarrow \mathcal{S}_1(X).$$

Since we have :

$$0 \longrightarrow L^{-1} \longrightarrow \mathcal{O}_X \longrightarrow \mathcal{O}_D \longrightarrow 0.$$

we see that $p_*\mathcal{O}_{\bar{D}} \cong \mathcal{O}_D$ from which we see that D is normal.

□

(0.4.2) **Remarks**. a) For D as above we have shown that there is a natural surjection $\mathcal{S}_i(X) \longrightarrow \mathcal{S}_i(D) \longrightarrow 0$,

b) Assume that X is also projective with $\dim X \geqslant 2$, L is ample and spanned, and A is generic. Then the proof of (0.4.1) shows that if Irr(A) is empty then $\mathrm{Irr}(X) \subset X-A$ is finite. Indeed the proof shows that if A is generic and Irr(A) is empty then $\mathrm{Irr}(X) \subset X-A$. Since A is ample, X-A is affine, and therefore Irr(X) is finite.

(0.4.3) **Lemma**. _Let X be an n dimensional projective variety. Let $\mathfrak{L}$ and L be line bundles on X that are spanned by global sections. Assume that $\mathbf{V} \subset \Gamma(L)$ spans L, that $\mathfrak{L}$ is big and that L is ample. Let x be a point of X. Let_

$$A_1, \ldots, A_{n-1} \text{ be general elements of } |\mathbf{V}-x|,$$

the linear system consisting of elements of $|\mathbf{V}|$ that contain x. Then the intersection of the A_i is a curve such that $\mathfrak{L}_C$ is ample.

Proof. To show that $\mathfrak{L}_C$ is ample it suffices to show that φ_C has finite fibres where $\varphi : X \longrightarrow \mathbb{P}_{\mathbb{C}}$ is the map associated to $\Gamma(\mathfrak{L})$. By induction it suffices to show that given a general $A \in |\mathbf{V}|$, then $\dim \varphi(A') = \dim A'$ for all irreducible components A' of A.

To see that these dimensions are equal note that the set of positive dimensional fibres of φ is an analytic set $\mathfrak{F}$. For $\dim \varphi(A')$ to be $< \dim A'$ it would be necessary that A' was a component of $\mathfrak{F}$. Since the base locus of $|\mathbf{V}-x|$ is finite, the set of elements of $|\mathbf{V}-x|$ that contain any of the finite number of irreducible components of $\mathfrak{F}$ is a proper analytic subset of $|\mathbf{V}-x|$. Therefore a general element of $|\mathbf{V}-x|$ contains no component of $\mathfrak{F}$.

$\square$

(0.5) **Reductions** ([So6]). Let L be an ample line bundle on a normal n dimensional projective variety X. A **reduction** (X',L') of the pair (X,L) is a pair (X',L') consisting of an ample line bundle L' on a normal projective variety X' such that :

a) there is a map $\pi : X' \longrightarrow X$ expressing X' as X with a finite set $F \subset \mathrm{reg}(X)$ blown up,

b) $L \cong \pi^* L' \otimes [\pi^{-1}(F)]^{-1}$ or equivalently $K_X \otimes L^{n-1} \cong \pi^*(K_{X'} \otimes L'^{n-1})$.

Note that the $A \in |L|$ that are smooth in a neighborhood of $\pi^{-1}(F)$ are in one to one correspondence with the $A' \in |L'|$ such that $F \subset \mathrm{reg}(A')$. The correspondence is gotten by sending such an A' to its proper transform A. If an $A \in |L|$ which is smooth in a neighborhood of $\pi^{-1}(F)$ exists then $L' = [\pi(A)]$.

We need theorem (1.1) of [(F+S)1]; see also [So5].

(0.6) **Theorem**. _Let_ L _be an ample line bundle on a smooth, connected, projective surface_ S. _If_ S _is a minimal non-negative Kodaira dimension surface then there exist arbitrarily large integers_ N _such that_ $\Gamma(K_S{}^n \otimes L^N)$ _spans_ $K_S{}^n \otimes L^N$ _for all_ $n \geq 0$. _Otherwise there exist arbitrarily large_ N _with the property that_ $\Gamma(K_S{}^n \otimes L^N)$ _spans_ $K_S{}^n \otimes L^N$ _for all non-negative integers_ $n \leq A$ _for some_ $A \geq 0$ _where either_ :

a) $A \geq N$ _and_ $K_S{}^A \otimes L^N$ _is not ample,_

b) $2A = N$ _and_ S _is a_ $\mathbb{P}^1$ _bundle_ $r : S \longrightarrow C$ _over a smooth curve_ C, _and_ $L_F \cong \mathcal{O}_{\mathbb{P}^1}(1)$ _for any fibre,_ F, _of_ r,

_or,_

c) $K_S{}^a \otimes L^N$ _is trivial for some_ $a > 0$.

§1 Spectral Values

In this section we define spectral values and prove the few general facts that we know about this notion. Though we can make sense of the definitions for more or less arbitrary pairs we restrict ourselves for simplicity to semi-ample and big line bundles on normal projective varieties.

Let L be a semi-ample and big line bundle on a normal projective variety X of dimension $n \geq 1$. We define the **spectral value**, $\sigma(X,L)$, of the pair (X,L) as the smallest real number τ such that given any fraction $p/q > \tau$

$$\Gamma(K_X^N \otimes L^{(n+1-p/q)N}) = 0 \text{ for all integers } N > 0 \text{ with } q \mid N.$$

Given a real number τ we define the spectral class $\mathfrak{X}_\tau$ as the set of all pairs (X,L) such that :

1) L is a semi-ample and big line bundle on a normal positive dimensional projective variety X,
2) $\sigma(X,L) < \tau$.

(1.1) **Lemma**. *Let L be a semi-ample and big line bundle on a normal projective variety X of dimension $n > 0$. If $p : \bar{X} \longrightarrow X$ be a resolution of singularities of X, then $\sigma(X,L) \geq \sigma(\bar{X},\sigma^*L)$.*

Proof. For all integers $N > 0$ we have the inclusion $0 \longrightarrow p_*(K_{\bar{X}}^N) \longrightarrow K_X^N$.

Therefore given any fraction p/q with $q \mid N$ we have

$$h^0(K_X^N \otimes L^{(n+1-p/q)N}) \geq h^0(K_{\bar{X}}^N \otimes (p^*L)^{(n+1-p/q)N}).$$

□

(1.2) **Theorem**. *If L is a semi-ample and big line bundle on a positive dimensional normal projective variety X, then $\sigma(X,L) \geq 0$.*

Proof. By the above lemma we can assume that X is smooth and projective.
If the theorem was false then $h^0(K_X \otimes L^N) = 0$ for $N = 1, \ldots, n+1$. By (0.2.1) and Serre duality the Hilbert polynomial $\chi(L^{-x})$ vanishes. This is absurd since the leading coefficient $[-c_1(L)^n]/n!$ of the polynomial is non-zero.

□

(1.3) **Lemma**. Let L be an ample line bundle on a normal projective variety X. Given any normal $A \in |L|$, $\sigma(X,L) \leq \sigma(A,L_A)$. In particular if $(A,L_A) \in \mathfrak{T}_\tau$ then $(X,L) \in \mathfrak{T}_\tau$.

Proof. Is is easy to check that there is a natural map :

$$K_X^N \otimes L^{(n+1-p/q)N} \longrightarrow K_A^N \otimes L_A^{(n-p/q)N}$$

which is onto over reg(X). Therefore if we have a not identically zero section s of $K_X^N \otimes L^{(n+1-p/q)}$ we will get a not identically zero section of $K_A^N \otimes L_A^{(n-p/q)N}$ unless s vanishes on A to some order k.

Since L is semi-ample there is some integer $M > 0$ such that L^M is spanned by global sections. Letting e be a general section of L^M and s_A be a section of L vanishing precisely on A we get the not identically zero section $s^M \otimes e^k/s_A^N$ of $K_X^{NM} \otimes L^{(n+1-p/q)NM}$ which restricts to a not identically zero section of $K_A^{NM} \otimes L_A^{(n-p/q)NM}$.

□

(1.4) **Lemma**. Let L be a semi-ample and big line bundle on a normal projective variety X. If q and p are integers such that :

$$h^0(K_X^N \otimes L^{(n+1-p/q)N}) \geq c \cdot N^n$$

for some constant $c > 0$ then $\sigma(X,L) > p/q$.

Proof. Let $\mathcal{L} = K_X^q \otimes L^{(n+1-p/q)q}$. Choose an integer $M > 0$ such that L^M has a section and let $A \in |L^M|$. To prove the lemma it suffices to show that $h^0(\mathcal{L}^N \otimes L^{-M}) > 0$ for some integer $N > 0$. Since $\dim A = n-1$, $h^0(\mathcal{L}_A^N) < c' \cdot N^{n-1}$ for some $c' > 0$. Considering :

$$0 \longrightarrow \mathcal{L}^N \otimes L^{-M} \longrightarrow \mathcal{L}^N \longrightarrow \mathcal{L}_A^N \longrightarrow 0$$

we see that $h^0(\mathcal{L}^N) \leq h^0(\mathcal{L}^N \otimes L^{-M}) + h^0(\mathcal{L}_A^N)$ which combined with the above inequality proves the lemma.

□

(1.5) **Question**. Assume that L is an ample and spanned line bundle on X, a smooth projective variety. Is $\sigma(X,L)$ rational ?

I know no example of a nef and big line bundle L on a normal projective variety X with $\sigma(X,L)$ irrational.

§ 2 On the Ascent of Numerical Effectivity

(2.0) In this section we show that under certain circumstances numerical effectivity and spannedness ascend off an ample divisor. These results let us carry out the inductions on which this paper is based.

The next theorem is the key new technical result of this paper.

(2.1) **Theorem**. Let X be a normal projective $\mathbb{Q}$-Gorenstein variety with both Irr(X). Let L be an ample line bundle on X and let $A \in |L|$. Assume that X is Gorenstein in a neighborhood of A. Let τ be a real number ≥ 2. If $(K_X + \tau L)_A$ is numerically effective then $K_X + \tau L$ is numerically effective.

Proof. Let

$$\bar{\tau} = \inf\{ t \in \mathbb{R} \mid K_X + tL \text{ is nef} \}.$$

If $\tau \geq \bar{\tau}$, then there is nothing to prove. Therefore we can assume that $\tau < \bar{\tau}$. Choose a rational number τ' such that $\tau < \tau' < \bar{\tau}$. For any $N > 0$ we can choose integers $p > q > N$ such that :

a) $2 \leq \tau < \tau' < p/q \leq \bar{\tau} < (p+1)/q$,
b) q is divisible by the index of X and the denominator of τ',
c) $p - \tau' q$ goes to infinity as N does.

Tensor the residue sequence with $K_X{}^q \otimes L^{p+1}$ to get

$$0 \longrightarrow K_X \otimes K_X{}^q \otimes L^{p+1} \longrightarrow K_X{}^{q+1} \otimes L^{p+2} \longrightarrow K_A{}^q \otimes L_A{}^{p-q+1} \otimes K_A \longrightarrow 0.$$

Note that by the numerical effectivity of $K_A + (\tau - 1)L_A$ and the ampleness of L it follows that $K_A + (\tau' - 1)L_A$ is ample. Therefore for N and hence q sufficiently large it follows that $K_A{}^q \otimes L_A{}^{(\tau'-1)q}$ is spanned by global sections. By c), $K_A \otimes L_A{}^{p-\tau q+1}$ is spanned by global sections for N sufficiently large. Thus

$$K_A{}^q \otimes L_A{}^{p-q+1} \otimes K_A \cong K_A{}^q \otimes L_A{}^{(\tau'-1)q} \otimes (K_A \otimes L_A{}^{p-\tau q+1})$$

is spanned by global sections. Further by a) it follows that $K_X{}^q \otimes L^{p+1}$ is ample and therefore by (0.3.1) and the above exact sequence it follows that

$$\Gamma(K_X{}^{q+1} \otimes L^{p+2}) \longrightarrow \Gamma(K_A{}^q \otimes L_A{}^{p-q+1} \otimes K_A) \longrightarrow 0.$$

Therefore $K_X{}^{q+1} \otimes L^{p+2}$ is spanned by global sections off a set that doesn't meet A. This set is of necessity finite since A is ample. Therefore $(q+1)K_X + (p+2)L$ is numerically effective. But since $p/q > \tau \geq 2$, $(p+2)/(q+1) - p/q = (2q-p)/[q(q+1)] < 0$.
Thus $p/q > \bar{\tau}$. This contradiction proves the theorem.

□

We need some results on the ascent of spannedness.

(2.2) **Theorem**. Let (X,L) be as in (2.1). Let $A \in |L|$ be normal Gorenstein with $\mathrm{Irr}(X)$ finite. Let τ be an integer ≥ 2. Then $K_X \otimes L^{\tau}$ is semi-ample if $K_A \otimes L_A{}^{\tau-1}$ is semi-ample.

Proof. Let r be the index of X. Consider the residue sequence :

$$0 \longrightarrow K_X \longrightarrow K_X \otimes L \longrightarrow K_A \longrightarrow 0.$$

Tensor this sequence with $(K_X{}^r \otimes L^{r\tau})^B \otimes L^{\tau-1}$ for an integer $B > 0$ to get :

$$0 \longrightarrow K_X \otimes (K_X{}^r \otimes L^{r\tau})^B \otimes L^{\tau-1} \longrightarrow K_X \otimes (K_X{}^r \otimes L^{r\tau}) \otimes L^{\tau} \longrightarrow (K_A \otimes L_A{}^{\tau-1})^{Br+1} \longrightarrow 0.$$

Choose B so that $(K_A \otimes L^{\tau-1})^{Br+1}$ is spanned by global sections. By (2.1) $K_X + \tau L$ is nef and so $(K_X{}^r \otimes L^{r\tau})^B \otimes L^{\tau-1}$ is ample. Therefore by (0.2.1)

$$\Gamma(K_X \otimes (K_X{}^r \otimes L^{r\tau}) \otimes L^{\tau}) \longrightarrow \Gamma((K_A \otimes L_A{}^{\tau-1})^{Br+1}) \longrightarrow 0$$

and we conclude that the analytic set where $K_X \otimes (K_X{}^r \otimes L^{r\tau}) \otimes L^{\tau}$ is not spanned by global sections is in X-A and therefore finite. Therefore $(K_X{}^r \otimes L^{r\tau})^B$ is spanned off a finite set for some integer $B > 0$. Therefore by the Fujita-Zariski theorem (see [F5]) the theorem is proven.

□

The next corollary will be used to recognize scrolls, quadrics, and Del Pezzo bundles.

(2.2.1) **Corollary**. Let L be an ample line bundle on a normal $\mathbb{Q}$-Gorenstein variety with $\mathrm{Irr}(X)$ finite. Assume that there is a normal Gorenstein $A \in |L|$ and a holomorphic surjection $\varphi : A \longrightarrow Y$ where $\dim A > \dim Y$, Y is normal, and φ has connected fibres. Assume further that $(K_A \otimes L_A{}^k) = \varphi^*\mathfrak{L}$ for some ample line bundle $\mathfrak{L}$ on Y. If $k \geq 1$ then φ extends to a holomorphic surjection $\psi : X \longrightarrow Y$ with $(K_X \otimes L^{k+1}) = \psi^*\mathfrak{L}$ and X is

<u>Gorenstein with</u> Irr(X) <u>empty</u>.

Proof. By theorem (2.2) $(K_X \otimes L^{k+1})^N$ is spanned for large N divisible by the index r of X. We can assume for N >> 0 that the map $\psi : X \longrightarrow \mathbb{P}_{\mathbb{C}}$ associated to $\Gamma((K_X \otimes L^{k+1})^N)$ has a normal image $Z = \psi(X)$ and ψ has connected fibres.

Since $(K_X \otimes L^{k+1})_A \cong K_A \otimes L_A{}^k$ we can conclude that we have a commutative diagram of holomorphic maps :

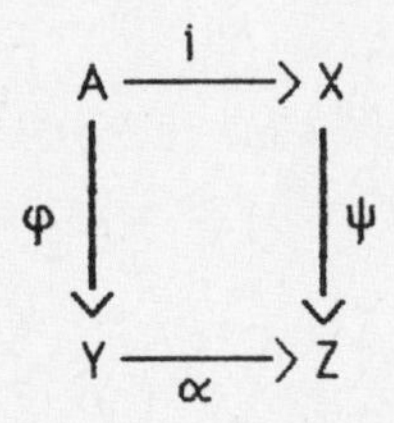

where i is the inclusion and α is finite to one. By (0.3.2) we conclude that $\psi(A) = \alpha(\varphi(A)) = \alpha(Y) = Z$. Since general fibres of both ψ and ψ_A are normal (e.g. by (0.4.1)) we can use dim A > dim Y and (0.2.3) to conclude that $\alpha \circ \varphi$ has connected fibres, i.e. that α is a biholomorphism and ψ is an extension of φ. Consider $\mathcal{S} = K_X \otimes L^{k+1} \otimes \psi^* \mathcal{I}^{-1}$. Since $\mathcal{S}_A \cong \mathcal{O}_A$ and since $\mathcal{S}^r$ is invertible we can conclude from (0.3.1) that $\mathcal{S} \cong \mathcal{O}_X$, i.e. that $K_X \otimes L^{k+1} \cong \psi^* \mathcal{I}$.

Note that K_X is invertible. Since $k \geqslant 1$ we know that $h^0(K_X \otimes L) = h^0(\psi^* \mathcal{I} \otimes L^{-k}) = 0$. By the proof of (0.2.2) and the fact that Irr(X) is finite we see that $p_* K_{\overline{X}} \cong K_X$ for any resolution $p\ \overline{X} \longrightarrow X$. Thus X has canonical singularities which are rational by the theorem of Elkik and Shepherd-Barron.

□

We need a variant of theorem (2.2).

(2.3) **Theorem**. <u>Let</u> X <u>be a normal</u>, n <u>dimensional</u>, <u>$\mathbb{Q}$-Gorenstein projective variety with</u> Irr(X) <u>finite</u>. <u>Assume that</u> L <u>and</u> $\mathcal{I}$ <u>are line bundles spanned by global sections</u>. <u>Assume that</u> L <u>is ample and</u> $\mathcal{I}$ <u>is big</u>. <u>Then there is an integer</u> M > 0 <u>such that</u> $K_X \otimes L^{n-1} \otimes \mathcal{I}^M$ <u>is semi-ample</u>.

Proof. By the Fujita-Zariski theorem [F5], it suffices to show that for M >> 0, $K_X \otimes L^{n-1} \otimes \mathcal{I}^M$ is spanned by global sections off the set of non-Gorenstein points of X.

Let $x \in X$ be a Gorenstein point of X and let $A_1, \ldots, A_{n-1}$ and C be as in lemma (0.4.3). Since the A_i are general it can be assumed without loss of generality that X is Gorenstein in a neighborhood of C. Let s_i for $1 \leqslant i \leqslant n-1$ be sections of L such that

$s_i^{-1}(0) = A_i$. Let $s = s_1 \oplus \cdots \oplus s_{n-1}$ be the induced section of $\mathbf{E}$, the direct sum of n-1 copies of L. Let :

$$0 \longrightarrow \mathbf{E} \longrightarrow \Lambda^2\mathbf{E} \longrightarrow \cdots \longrightarrow \Lambda^{n-1}\mathbf{E} \longrightarrow \Lambda^{n-1}\mathbf{E}_C \longrightarrow 0$$

be the induced Koszul complex. Let Θ denote the tensor product of this complex with $K_X \otimes \mathcal{L}^M$. Since $\mathcal{L}_C$ is ample, it follows that for $M >> 0$, $K_X \otimes \mathcal{L}^M \otimes (\Lambda^{n-1}\mathbf{E})_C$ is spanned by global sectons. Since $K_X \otimes \mathcal{L}^M \otimes \Lambda^{n-1}\mathbf{E} = K_X \otimes \mathcal{L}^M \otimes L^{n-1}$, we will done by the hypercohomology spectral sequence of Θ if we show that $H^i(K_X \otimes \mathcal{L}^M \otimes \Lambda^{n-i-1}\mathbf{E}) = 0$ for $i > 0$ and $M > 0$. Since Irr(X) is finite, and $\Lambda^{n-i-1}\mathbf{E}$ is a direct sum of copies of L^{n-i-1} for $i \leqslant n-1$, this vanishing is a direct consequence of (0.2.1).

□

§ 3 Special Varieties

(3.0) There are various classes of special varieties such as scrolls, quadric bundles, Del Pezzo bundles, and generalized cones which play a key role in this paper. In this section we collect the necessary background material about these varieties and prove those facts for which we know no good reference.

By a **quadric** we mean a pair $(\mathbb{Q}, \mathcal{O}_{\mathbb{Q}}(1))$ where $\mathbb{Q}$ is a degree 2 hypersurface of $\mathbb{P}^{n+1}$ and $\mathcal{O}_{\mathbb{Q}}(1) \cong \mathcal{O}_{\mathbb{P}}(1)_{\mathbb{Q}}$. By a **Del Pezzo variety** we mean a pair (X,L) consisting of a normal projective variety, X, and L an ample line bundle on X which is spanned by global sections and satisfies $K_X \otimes L^{\dim X - 1} \cong \mathcal{O}_X$.

Let L be an ample and spanned line bundle on a normal $\mathbb{Q}$-Gorenstein projective variety, X, of index r. We say that (X,L) is a **scroll** (respectively a **quadric bundle**, respectively a **Del Pezzo bundle**) if there exists a surjective holomorphic map $\varphi : X \longrightarrow Y$ onto a normal variety Y such that :

a) φ has connected fibres and $\dim X > \dim Y$,
b) $K_X^r \otimes L^{Tr} = \varphi^*\mathcal{L}$ where $\mathcal{L}$ is an ample line bundle on Y and $T = \dim X - \dim Y + 1$ (respectively $T = \dim X - \dim Y$, respectively $T = \dim X - \dim Y - 1$).

(3.1) **Theorem**. Let (X,L) be a scroll. Let $\varphi : X \longrightarrow Y$ and $\mathcal{L}$ be as above and set $k = \dim X - \dim Y$. The induced map from $\varphi^{-1}(U)$ to U is a $\mathbb{P}^k$ bundle over the Zariski open set U of $y \in Y$ with $\dim \varphi^{-1}(y) - k$. For a generic fibre F of φ, $L_F \cong \mathcal{O}_{\mathbb{P}}(1)$. Further $\varphi^{-1}(U)$ -

$\mathbb{P}(\varphi^* L_U)$.

Proof. By the upper semi-continuity in the Zariski topology of dimension of the fibres of a holomorphic map it follows that U is Zariski open.

Choose a very ample line bundle M on Y. Using (0.4.1) and dim Y generic elements of $|\varphi^* M| = \varphi^*|M|$ it can be seen that for a generic fibre F of φ :

a) F is normal and $K_{X,F} \cong K_F$,
b) $K_F \otimes L_F^{k+1} \cong \mathcal{O}_F$.

Assume that $k = 1$. It is clear that $(F,L_F) \cong (\mathbb{P}^1, \mathcal{O}_{\mathbb{P}}(1))$. We also see that $L \cdot \varphi^{-1}(y) = 1$ for any $y \in U$. Thus by ampleness we see that all such fibres are reduced and irreducible. Let $y \in U$ and choose 2 general elements $A_1, A_2 \in |L|$ such that $A_1 \cap A_2 \cap \varphi^{-1}(y)$ is empty; we can do this since $\varphi^{-1}(y)$ is one dimensional. Let s_1, s_2 be the sections of L corresponding to A_1, A_2. Using s_1, s_2 we get a holomorphic map $\alpha : \varphi^{-1}(V) \longrightarrow \mathbb{P}^1$ for some neighborhood V of y. Consider the map $(\alpha,\varphi) : \varphi^{-1}(V) \longrightarrow \mathbb{P}^1 x V$. This map is finite to one and generically one to one by construction and therefore by Zariski's main theorem the map is a biholomorphism.

Assume now that $k > 1$. Let $\bar{y} \in U$. Choose a general $A \in |L|$ such that $\dim \varphi_A{}^{-1}(y) = k-1$ for y in some neighborhood V of $\bar{y}$. By induction on dim X - dim Y, $A \cap \varphi^{-1}(V)$ is a $\mathbb{P}^{k-1}$ bundle over V with repect to the map induced by φ. Since $\mathcal{O}_{\mathbb{P}}(1)$ is spanned by no fewer than all its sections and since L is spanned we can choose sections $z_0, \ldots, z_{k-1}$ of L that span L over $\varphi^{-1}(V)$ where V is a possibly smaller neighborhood of $\bar{y}$. Let s be the section corresponding to A. We get a map $\alpha : \varphi^{-1}(V) \longrightarrow \mathbb{P}^k$ by using $\{z_0, \ldots, z_{k-1}, s\}$. Thus we get a map $(\alpha,\varphi) : \varphi^{-1}(V) \longrightarrow \mathbb{P}^k x V$. We will be done by the same reasoning as in the first paragraph if we know that (α,φ) is generically one to one. To see this we must merely show that $\alpha_F : F \longrightarrow \mathbb{P}^k$ is one to one for a generic fibre $F \subset \varphi^{-1}(V)$. This is easy since α_F is finite to one by ampleness and $c_1(L_F)^k[F] = c_1(L_{F \cap A})^{k-1}[F \cap A] = 1$. In particular $(F,L_F) \cong (\mathbb{P}^k, \mathcal{O}_{\mathbb{P}}(1))$.

□

By a slight variant of the above argument we can also prove.

(3.2) **Theorem**. <u>Let</u> (X,L) <u>be a quadric bundle</u>. <u>Let</u> $\varphi : X \longrightarrow Y$, <u>and let</u> Σ <u>be as in the definition above</u>. <u>Let</u> $k = \dim X - \dim Y$. <u>Let</u> $U \subset Y$ <u>be the Zariski open set of</u> $Y \in Y$ <u>with</u> $\dim \varphi^{-1}(y) = k$. <u>Then the natural meromorphic map</u> $\varphi^{-1}(U) \longrightarrow \mathbb{P}(\varphi^* L)$ <u>is a holomorphic embedding</u>. <u>The general fibre</u> F <u>of</u> φ <u>is a quadric</u>.

(3.3) **Theorem**. Let (X,L) be a scroll with $\varphi : X \longrightarrow Y$ and $\mathcal{L}$ as in the definition of a scroll and $k = \dim X - \dim Y$. Let $Z \subset Y$ be the union of $\mathrm{Sing}(Y)$ and those points $y \in Y$ such that $\dim \varphi^{-1}(y) > k$. If X is smooth and $k > 1$ then $\mathrm{cod}\, Z \geqslant 3$, e.g. if $\dim Y \leqslant 2$ then φ is a $\mathbb{P}^k$ bundle over a smooth base.

Proof. Choose a very ample line bundle M on Y. Using (0.4.1) and generic elements of $|\varphi^*M| = \varphi^*|M|$ it can be assumed without loss of generality that Z is finite. Thus if the theorem is shown for $\dim Y = \mathrm{cod}\, Z \leqslant 2$ then we will be done.

If $\dim Y = 1$ then all fibres are equal dimensional and the lemma follows from (3.1). Therefore we can assume that $\dim Y = 2$.

Choosing k general elements $A_1, \; , \; , A_k$ of $|L|$ and intersecting we get a smooth surface S such that $\varphi_S : S \longrightarrow Y$ is birational and $K_S \otimes L = \varphi_S{}^*\mathcal{L}$ for an ample line bundle $\mathcal{L}$ on Y. Using the argument of [So2] about this situation we see that Y is smooth and $\varphi_S : S \longrightarrow Y$ expresses S as Y with a finite set blown up. Assume that a fibre F of φ is of dimension $> k$ and hence a divisor. Using an induction down to S we can deduce that $(F,L_F) \cong (\mathbb{P}^{k+1},\mathcal{O}_{\mathbb{P}}(1))$. Since $(K_X \otimes L^{k+1})_F \cong \mathcal{O}_F$ we conclude that $N_F \cong \mathcal{O}_F(-1)$. Therefore there exists a map $q : X \longrightarrow X'$ blowing down the divisor F, and φ factors $\varphi = s \circ q$ where $s : X \longrightarrow X'$. Since $s^{-1}(\varphi(F))$ is a point and the general fibre of s is a positive k dimensional variety we get a contradiction to the upper semi-continuity of fibre dimensions of a holomorphic map.

□

The following theorem shows that the condition $\mathrm{cod}\, Z \geqslant 3$ cannot be improved.

(3.3.2) **Example**. Let $\mathbf{P} = \mathbb{P}(\mathcal{O}_{\mathbb{P}^3}(1) \oplus \mathcal{O}_{\mathbb{P}^3}(1) \oplus \mathcal{O}_{\mathbb{P}^3}(1))$. Let s be the section of the tautological bundle ξ on $\mathbf{P}$ that corresponds to the section $z_0 \oplus z_1 \oplus z_2$ where z_1, z_2, z_3 are homogeneous coordinates considered as sections of $\mathcal{O}_{\mathbb{P}^3}(1)$. Let $X = s^{-1}(0) \subset \mathbf{P}$. Then (X,ξ_X) is a scroll with precisely 1 two dimensional fibre, the variety Σ over $(0,0,0,1) \in \mathbb{P}^3$. Note that $N_\Sigma \cong T_{\mathbb{P}^2}{}^*(-1)$ which shows why the argument (3.3) fails.

Theorem (3.3) fails if X is only Gorenstein. The following example illustrates what can happen in the 3 dimensional case.

(3.3.3) **Example**. Choose a smooth projective surface S'. Let $\pi : S \longrightarrow S'$ be the blowup of S' at one point x and let $E = \pi^{-1}(x)$. Let $\mathcal{L}$ be a very ample line bundle on S'. $\mathcal{L}' = \pi^*\mathcal{L}^2 \otimes [E]^{-1}$ is very ample on S. Let $X' = \mathbb{P}(\mathcal{L}' \oplus \pi^*\mathcal{L})$. We have the diagram

$$\begin{array}{ccc} X' & \xrightarrow{\overline{\pi}} & S \\ \varphi \downarrow & & \downarrow \pi \\ X & \xrightarrow{p} & S' \end{array}$$

Here $\varphi : X' \longrightarrow \mathbb{P}_{\mathbb{C}}$ with $X = \varphi(X')$ is the map associated to $\Gamma(\xi)$ where ξ is the tautological line bundle on X', and X is precisely X' with the curve C over E corresponding to $(\mathcal{L}' \oplus \pi^*\mathcal{L})_E \longrightarrow \mathcal{O}_E \longrightarrow 0$ contracted. Further $\overline{\pi}$ is the tautological projection and p is the holomorphic map that sends $x \in X$ to $\pi(\overline{\pi}(\varphi^{-1}(x)))$. Note that $N_C \cong \mathcal{O}_{\mathbb{P}^1}(-1) \oplus \mathcal{O}_{\mathbb{P}^1}(-1)$. Therefore by the adjunction formula $K_{X',C} \cong \mathcal{O}_C$ and we can conclude that X is Gorenstein. Let $L = \mathcal{O}_{\mathbb{P}^1}(1)_X$ and note that

$$\varphi^*(K_X \otimes L^2) \cong K_{X'} \otimes \xi^2 \cong \overline{\pi}^*(K_S \otimes \mathcal{L}' \otimes \pi^*\mathcal{L}) \cong \overline{\pi}^*(\pi^*(K_{S'} \otimes \mathcal{L}^2)).$$

Using the fact that $K_{S'} \otimes \mathcal{L}^2$ is ample for most choices of $(S',\mathcal{L})$ we have an example of a scroll (X,L) with projection $p : X \longrightarrow S'$, $\mathbb{P}^2$ as exceptional fibre and $\mathbb{P}^1$ as generic fibre.

(3.4) **Generalized Cones**. Let L be a very ample line bundle on a pure dimenstional projective variety X of dimension n. Let

$$E = \underbrace{\mathcal{O}_X \oplus \cdots \oplus \mathcal{O}_X}_{N-n \text{ copies}}$$

where N is an integer $\geq n$. Let $\mathbf{C} = \mathbb{P}(E \oplus L)$ and let ξ denote the tautological line bundle on $\mathbf{C}$. Let $\varphi : \mathbf{C} \longrightarrow \mathbb{P}_{\mathbb{C}}$ denote the map associated to $\Gamma(\xi)$. We denote $\varphi(\mathbf{C})$ by $\mathbf{C_N}(X,L)$ and we denote the restriction of $\mathcal{O}_{\mathbb{P}}(1)$ to $\mathbf{C_N}(X,L)$ by ξ_L. We call $(\mathbf{C_N}(X,L),\xi_L)$, or $\mathbf{C_N}(X,L)$ for short, the **generalized cone of dimension N on (X,L)**. If N = n then $\mathbf{C_N}(X,L) = X$ and $\xi_L = L$. If N > n we say that the generalized cone $\mathbf{C_N}(X,L)$ is non-degenerate. We call (X,L) the **base** of the cone and $(\varphi(D),\xi_{L,\varphi(D)})$ the **nappe** of the cone where $D \cong \mathbb{P}(E)$ is the variety on $\mathbf{C}$ determined by $E \otimes L \longrightarrow E \longrightarrow 0$.

To see what $\mathbf{C_N}(X,L)$ looks like when N > n, note that we have the divisor $D \cong X \times \mathbb{P}^{N-n+1}$ defined on $\mathbf{C_N}(X,L)$. The map φ is a biholomorphism on $\mathbf{C_N}(X,L)$-D and on D the map φ_D is the product projection $X \times \mathbb{P}^{N-n-1} \longrightarrow \mathbb{P}^{N-n-1}$

(3.4.1) **Lemma**. *Let L be a very ample line bundle on a $\mathbb{Q}$-Gorenstein normal projective variey X. Then the generalized cone $\mathbf{C_N}(X,L)$ is $\mathbb{Q}$-Gorenstein for some N > n if and only if there is a positive integer a and an integer b such that $K_X{}^a \cong L^{-b}$.*

Proof. Note that away from the nappe of $\mathbf{C_N}(X,L)$, the cone is locally a product of X with a manifold of dimension N-n. Therefore the complement of the nappe of the cone is $\mathbb{Q}$-Gorenstein. We must therefore only investigate a neighborhood of the nappe of the cone.

Let Y be the normal variety obtained from L^{-1} by collapsing the zero section of L^{-1} to a point. If $\Gamma(L)$ embeds X in $\mathbb{P}^K$, then Y is simply the cone over X in $\mathbb{C}^{K+1}$. In a neighborhood of the nappe of $\mathbf{C_N}(X,L)$, the cone $\mathbf{C_N}(X,L)$ is locally a product of $\mathbb{C}^{N-n-1}$ and a neighborhood in Y of the singularity of Y. Thus we must only investigate Y. Note that, $\mathbf{K}$, the canonical bundle of L^{-1} is isomorphic to $\pi^*(K_X \otimes L)$ wherer $\pi : L^{-1} \longrightarrow X$ is the tautological projection and $\pi^*L^{-1} \cong [X]$ where X is considered as the zero section of L^{-1}. It is easy to see that K_Y is $\mathbb{Q}$-Gorenstein if and only if there is an integer $a > 0$ and an integer c such that in a neighborhood of X $\mathbf{K}^a \cong [X]^c$, i.e. that $\pi^*(K_X{}^a \otimes L^{-b})$ is trivial in a neighborhood of X for some integer b. It is straightforward to check that $\pi^*(K_X{}^a \otimes L^{-b})$ is trivial in a neighborhood of X if and only if $K_X{}^a \otimes L^{-b}$ is trivial on X.

□

We will need a certain structure theorem of Fujita [F1]. In our terminology his result (which is actually a little more general) is the following.

(3.5) **Fujita's Structure Theorem [F1].** *Let L be an ample and spanned line bundle on a normal, n dimensional, projective variety, X. Assume that g(L)) = 0 where*

$$2g(L)-2 = (K_X + (n-1)L) \cdot \underbrace{L \cdot \cdots L}_{n-1 \text{ times}}.$$

Then L is very ample and Irr(X) is empty. Further either (X,L) is $(\mathbb{P}^n, \mathcal{O}_{\mathbb{P}}(1))$ or a quadric or $n \geq 2$ and (X,L) is a (possibly degenerate) generalized cone on one of the following pairs:

a) *a scroll over a smooth curve of genus 0,*
b) $(\mathbb{P}^2, \mathcal{O}_{\mathbb{P}}(2))$,
c) $(\mathbb{P}^1, \mathcal{O}_{\mathbb{P}}(e))$ *with* $e \geq 3$.

(3.5.1) **Corollary.** *Let (X,L) be as in (3.5). Then X is $\mathbb{Q}$-Gorenstein except if (X,L) is a non-degenerate generalized cone on a scroll.*

Proof. By (3.4.1) the only possible non $\mathbb{Q}$-Gorenstein varieties on the list in (3.5) are the non-degenerate generalized cones on the scrolls in (3.5a). Let (X,L) be such a scroll. Here $X = \mathbb{P}(E)$ for some ample vector bundle on $\mathbb{P}^1$. In this case we have $K_{\mathbb{P}(E)} = L^{-\mathrm{rank}E} \otimes \pi^*(K_{\mathbb{P}^1} \otimes \det E)$ where $\pi : X \longrightarrow \mathbb{P}^1$ is the projection quaranteed by the definition of a scroll and $E = \pi_* L$. Using (3.4.1) we see that a non-degenerate generalized cone on (X,L) is $\mathbb{Q}$-Gorenstein if and only if $K_{\mathbb{P}^1} \otimes \det E \cong \mathcal{O}_{\mathbb{P}^1}$, i.e. if and only if $\det E = \mathcal{O}_{\mathbb{P}}(2)$. Since E is ample of rank ≥ 2 this implies tha $E \cong \mathcal{O}_{\mathbb{P}^1}(1) \oplus \mathcal{O}_{\mathbb{P}^1}(1)$. But in this case (X,L) is a quadric and not a scroll.

□

(3.5.2) **Remarks**. Let **C** denote $\mathbf{C_N}(X,L)$. If (X,L) is $(\mathbb{P}^n, \mathcal{O}_{\mathbb{P}}(1))$ then **C** is just $(\mathbb{P}^N, \mathcal{O}_{\mathbb{P}}(1))$. If (X,L) is a quadric, then so is **C**. If $(X,L) = (\mathbb{P}^2, \mathcal{O}_{\mathbb{P}}(2))$ then $K_{\mathbf{C}}{}^2 \cong \xi_L$ and if $(X,L) = (\mathbb{P}^1, \mathcal{O}_{\mathbb{P}^1}(e))$ with $e \geq 3$ then $K_{\mathbf{C}}{}^e \cong \xi^{-(eN+2-e)}$. If **C** is non degenerate then the index is 2 and e respectively in the last 2 cases.

The following theorem which supplements the above result will play a key role in the next section.

(3.6) **Theorem**. Let L be an ample and spanned line bundle on a normal Cohen-Macaulay projective variety X of dimension $n \geq 2$. The following conditions are equivalent :

a) $h^1(\mathcal{O}_X) = g(L)$,

b) $h^{n-1}(K_X) = g(L)$,

c) (X,L) is as in (3.5) or (X,L) is a scroll over a smooth curve of genus g(L).

Proof. By Serre duality a) and b) are equivalent. It is easy to check that c) implies a). Therefore we can assume that a) and b) are true. Our goal is to prove that c) is true. By (3.5) we can assume without loss of generality that $g(L) > 0$.

Assume first that $n = 2$. Choose a general $D \in |L|$ and let :

$$0 \longrightarrow K_X \longrightarrow K_X \otimes L \longrightarrow K_D \longrightarrow 0$$

be the residue sequence. This makes sense since D is smooth and misses Sing(X). By (0.2.1b), $h^1(K_X \otimes L) = 0$. Thus since $g(L) = h^0(K_D) = h^1(K_X)$ it follows that the map $\Gamma(K_X \otimes L) \longrightarrow \Gamma(K_D)$ is the zero map. Since L is spanned and D is general it follows that $\Gamma(K_X \otimes L)$ is 0. By (0.2.2) it follows that X has rational singularities. By (0.3.3) we have an Albanese map $\beta : X \longrightarrow \mathrm{ALB}(X)$ with $\dim \mathrm{ALB}(X) = h^1(\mathcal{O}_X) > 0$ and β^* giving an isomorphism of $H^1(\mathrm{ALB}(X), \mathcal{O}_{\mathrm{ALB}(X)})$ and $H^1(\mathcal{O}_X)$. Since X has rational singularities $\Gamma(K_X) \cong \Gamma(K_{\bar{X}})$ where $p : \bar{X} \longrightarrow X$ is a desingularization of X. Since $h^0(K_X) \leq h^0(K_X \otimes L) = 0$ it follows that $\beta(X)$ is a smooth curve C of genus $g(L) = h^1(\mathcal{O}_X)$ and β has connected fibres. Let $\pi : X \longrightarrow C$ denote this map.

Note that $\pi_D : D \longrightarrow C$ is a biholomorphism for general $D \in |L|$. If $g(L) > 1$, then this is an easy consequence of Hurwitz's formula. If $g(L) = 1$ then Hurwitz's formula implies that π_D is an unramified cover. Let X_D be the fibre product of π and π_D. Let $q: X_D \longrightarrow X$ be the induced unramified covering map. D induces a section σ of $X_D \longrightarrow D$. If π_D is not one to one there is another disjoint curve σ' so that $q^{-1}(D) = \sigma + \sigma'$. Since $q^{-1}(D)$ is ample this would give a contradiction to the fact that ample divisors on normal varieties of dimension at least 2 are connected (see(0.2.3)). Therefore π_D is one to one and $L \cdot f = 1$ for any fibre f of π. Thus every fibre is irreducible and reduced. Since L is also ample and spanned it follows that each fibre is a smooth $\mathbb{P}^1$.

The argument in the proof of (3.1) shows that $X \longrightarrow C$ is a $\mathbb{P}^1$ bundle. Finally $(K_X \otimes L^2)_f$ is trivial for all fibres f and therefore we conclude that $K_X \otimes L^2 = \pi^* \mathfrak{L}$ for some line bundle $\mathfrak{L}$ on C. To see that $\mathfrak{L}$ is ample we must show that $\deg(\mathfrak{L}) > 0$ or equivalently that $(K_X + L) \cdot L > 0$. To see this note that :

$$(K_X + 2L) \cdot L = (K_X + L) \cdot L + L \cdot L = 2g(L) - 2 + L \cdot L \geq L \cdot L > 0.$$

As an induction hypothesis we can assume that the assertion **a) implies c)** has been proven for $\dim X < n$ and $n \geq 3$.

Since X is normal, Sing(X) has codimension at least 2. By inductively choosing generic $A_i \in |L|$ we can find $n - 1$ divisors $A_1, \ldots, A_{n-1}$ such that :

1) $\mathfrak{D}_0 = X$ and for i from 1 to n-1 $\mathfrak{D}_i = A_1 \cap \ldots \cap A_i$ is normal and irreducible and $\dim \operatorname{Irr}(\mathfrak{D}_i) \leq \max \{0, \dim \operatorname{Irr}(\mathfrak{D}_{i-1}) - 1\}$,

2) $\mathfrak{D}_{n-1}$ is a smooth curve of genus g(L).

Consider the exact sequences :

$$0 \longrightarrow [\mathfrak{D}_i]^{-1} \longrightarrow \mathcal{O}_{i-1} \longrightarrow \mathcal{O}_i \longrightarrow 0$$

where $i = 1, \ldots, n-1$ and $\mathcal{O}_i$ is the structure sheaf of $\mathfrak{D}_i$. By (0.2.4) we have :

*) $$h^1(\mathcal{O}_X) \leq h^1(\mathcal{O}_1) \leq \ldots \leq h^1(\mathcal{O}_{n-1}) = g(L).$$

Since $h^1(\mathcal{O}_X) = g(L)$, $h^1(\mathcal{O}_1) = g(L)$ and by the induction hypothesis $A_1 = \mathfrak{D}_1$ is $\mathbb{P}(E)$ for a rank n-1 ample vector bundle E on C with map $\overline{\pi} : A_1 \longrightarrow C$ and $K_{\mathbb{P}(E)} \otimes L_{\mathbb{P}(E)}{}^{n-1} = \overline{\pi}^* \mathfrak{L}$ for an ample line bundle $\mathfrak{L}$ on C. Since $h^0(K_A) = 0$ for general $A \in |L|$ we conclude as before that $h^0(K_X \otimes L) = 0$. Therefore X has rational singularities by (0.2.2). Using the Albanese mapping of (0.3.3) we get an extension $\pi : X \longrightarrow C$ of $\overline{\pi}$.

We see that since A_1 is smooth it follows that π is of maximal rank off a finite set $F \subset X-A_1$ which includes the singular set of X. Note that since $\pi : A_1 \longrightarrow C$ is a $\mathbb{P}^{n-1}$ bundle we can conclude that any fibre f of π is irreducible and reduced.

Let f be any fibre of π. It is a Cartier divisor since π is a map to a smooth curve. Since $f \cap F$ is finite we conclude that f is a smooth Cartier divisor off a finite set. By ([A+K], (2.14)) we conclude that f is normal. Using ([F1], (2.1d)) we see that $(f, L_f) = (\mathbb{P}^{n-1}, \mathcal{O}_{\mathbb{P}}(1))$. Arguing as before we conclude that $X = \mathbb{P}(\pi_* L)$ and $K_X \otimes L^n \cong \pi^* \mathfrak{L}$ for an ample line bundle $\mathfrak{L}$ on C.

□

(3.6.1) **Corollary**. Let L be an ample and spanned line bundle on a normal Cohen-Macaulay projective variety X of dimension $n \geq 2$. Assume that Irr(X) is finite. Then a), b), or c) of (3.6) are equivalent to either :

α) $(X,L) \in \mathcal{K}_2$ where $\mathcal{K}_2$ is as in section 1, i.e. $h^0((K_X \otimes L^{n-1})^N) = 0$ for all integers $N > 0$,

or,

β) $h^0(K_X \otimes L^{n-1}) = 0$.

Proof. To see that (3.6c) implies α) above, assume otherwise. If $h^0((K_X \otimes L^{n-1})^{N'}) \neq 0$ for some $N' > 0$. Then it follows that $g(L) \geq 1$ since

$$2g(L)-2 = (K_X + (n-1)L) \cdot \underbrace{L \cdot \cdots \cdot L}_{n-1 \text{ times}} \geq 0.$$

By (3.6) we see that (X,L) is a scroll over a smooth curve C. But $(K_X \otimes L^n)_F \cong \mathcal{O}_F$ on fibres, F, of the scroll $X \longrightarrow C$ which implies that $(K_X \otimes L^{n-1})_F$ is negative for all such F. Thus $h^0((K_X \otimes L^{n-1})^N) = 0$ for all integers $N > 0$.

Clearly α) implies β). Therefore we only need to show that β) implies (3.6a). Let $A \in |L|$ be chosen by (0.4.1) so that A is normal and Irr(X) is empty. Using the residue sequence, $h^0(K_X \otimes L^{n-1}) = 0$, and (0.2.1b), we see that $h^0(K_A) = h^1(K_X)$. If $n = 2$ this implies that $g(L) = h^1(\mathcal{O}_X)$ by Serre duality. Therefore we can assume by induction that $g(L) = h^1(\mathcal{O}_A)$ where $n \geq 3$. By the residue sequence and (0.2.1b) $h^{n-1}(K_X) = h^{n-2}(K_A)$. Since $g(L) = h^1(\mathcal{O}_A)$ we conclude by Serre duality that :

$$h^1(\mathcal{O}_X) = h^{n-1}(K_X) = h^{n-2}(K_A) = h^1(\mathcal{O}_A) = g(L).$$

□

§ 4 The Spectral Class $\mathfrak{X}_2$

(4.0) Throughout this section L is an ample and spanned line bundle on a normal $\mathbb{Q}$-Gorenstein projective variety X of dimension n where $n \geq 2$ unless otherwise specified. It is further assumed that Irr(X) and the locus of non Gorenstein points are finite.

(4.1) **Theorem**. Let X and L be as in (4.0). Then any of the following conditions are equivalent and imply that X is Cohen-Macaulay with only rational singularities:

a) $h^1(\mathcal{O}_X) = g(L)$,
b) $h^{n-1}(K_X) = g(L)$,
c) (X,L) is as in (3.5) or (X,L) is a scroll over a smooth curve of genus g(L),
d) $(X,L) \in \mathfrak{X}_2$, i.e. $h^0((K_X \otimes L^{n-1})^N) = 0$ for all integers $N > 0$,
e) $h^0(K_X \otimes L^{n-1}) = 0$.

Proof. If $g(L) = 0$ the the theorem follows from (3.5). Therefore it can be assumed without loss of generality that $g(L) > 1$.

If $n = 2$, then X is Cohen-Macaulay and the result follows from (3.6) and (3.6.1). Therefore we can assume as an induction hypothesis that the theorem has been proven in dimensions $< n$ where $n \geq 3$.

Choose a normal $A \in |L|$ with Irr(A) empty. By the same reasoning as used in (3.6) and (3.6.1) it can be seen that any of the conditions a) through e) is inherited by (A, L_A). Therefore by the induction hypothesis and the assumption that $g(L) > 0$ we see from that (A, L_A) is a scroll over a smooth curve C. By (2.2.1), (X,L) is a scroll over C and hence X is smooth. The result now follows from (3.6) and (3.6.1).

□

(4.1.1) **Corollary**. (X,L) be as in (4.0). $(A, L_A) \in \mathfrak{X}_2$, if and only if $(X,L) \in \mathfrak{X}_2$.

Proof. Use lemma (1.3) and the above theorem.

□

(4.2) **Theorem**. Let (X,L) be as in (4.0). The following are equivalent:

a) $K_X + (n-1)L$ is nef,
b) $(X,L) \notin \mathfrak{X}_2$.

Proof. Using (4.1) it is a direct calculation that a) implies b). Therefore it suffices to show that if $K_X + (n-1)L$ is not nef then $(X,L) \in \mathcal{X}_2$.

If dim $X = 2$ this follows immediately from Sakai's results [Sa3]. Therefore we can assume that $n \geq 3$ and by inductive hypothesis that the result has been proven for dimensions $< n$.

Let A be any normal irreducible element of $|L|$; such elements exist by (0.4.1) and (0.2.3). Since $K_X + (n-1)L$ is not nef and since $n \geq 3$, it follows from theorem (1.1) that $(K_X + (n-1)L)_A \cong K_A + (n-2)L_A$ is not nef. Therefore by the induction hypothesis it follows that $(A,L_A) \in \mathcal{X}_2$.

□

(4.2.1) **Corollary.** Let (X,L) be as in (4.0). Then either :

a) $(X,L) \cong (\mathbb{P}^n, \mathcal{O}_{\mathbb{P}^n}(1))$,

or

b) $K_X \otimes L^n$ is semi-ample.

Proof. If $(X,L) \in \mathcal{X}_2$ then we can use theorem (4.1) to check this by inspection. If not then by (4.2) we conclude that $K_X \otimes L^{n-1}$ is nef. From this we conclude that $K_X \otimes L^n$ is ample and hence semi-ample.

□

(4.3) **Remark.** If $K_X \otimes L^n$ is semi-ample but not ample then either $K_X \otimes L^n$ is trivial and (X,L) is a quadric, or (X,L) is a smooth scroll over a curve. This can be seen by using (4.1) or by simply studying the map associated to $\Gamma((K_X \otimes L^n)^t)$ for $t >> 0$ and divisible by the index of X.

Let $\log_2(X,L) = -\infty$ if $(X,L) \in \mathcal{X}_2$ and otherwise let it be the dimension of the image of X under the map associated to $\Gamma((K_X \otimes L^{n-1})^{rM})$ for integers $M >> 0$.

The following result is due to F. Sakai when dim $X = 2$.

(4.4) **Theorem.** Let (X,L) be as above as in (4.0) with $n \geq 1$. The following are equivalent :

a) $K_X \otimes L^{n-1}$ is nef,

b) $K_X \otimes L^{n-1}$ is semi-ample,

We have the following table.

$\log_2(X,L)$	**Structure of (X,L)**	**Sing(X)**
$-\infty$	Class $\mathcal{X}_2$	Rational Singularities in particular Cohen-Macaulay
0	$K_X \cong L^{-(n-1)}$ generalized Del Pezzo varieties, cf. [F7]	Gorenstein rational singularities if $n \geq 3$
1	Quadric bundle over a curve	Rational double points; in particular local complete intersection
$2 < n$	Scroll over a surface	Rational Singularities Gorenstein
> 2 and $< n$	don't occur	

<u>**Proof**</u>. Clearly b) implies a). Therefore we have to verify that a) implies b) and that the table is correct. We will do both of these together by induction. We have the properties :

A_n) a) implies b) for dim $X = n$.
B_n) the table is correct for dim $X = n$.

It is trivial that A_1 and B_1 are true. Further note that A_2 and B_2 are true by the main theorem of Sakai in [Sa3].

(4.4.1) <u>**Lemma**</u>. A_n <u>implies</u> A_{n+1} <u>for</u> $n \geq 2$.

<u>**Proof**</u>. By (0.4.1) we can choose a general element A of $|L|$ which is normal. We have $(K_X \otimes L)_A \cong K_A$. By hypothesis dim $X = n+1$ and $K_A \otimes L_A^{n-1}$ is nef. If $K_A \otimes L_A^{n-1}$ is semi-ample then $K_X \otimes L_X^n$ is semi-ample by theorem (2.2).

□

(4.4.2) <u>**Lemma**</u>. B_{n-1} <u>and</u> A_n <u>imply</u> B_n <u>for</u> $n \geq 3$.

<u>**Proof**</u>. Clearly $\log_1(X,L) = -\infty$ if and only if $(X,L) \in \mathcal{X}_2$. Therefore by (4.2) it can be assumed that $K_X \otimes L^{n-1}$ is nef. Since we are assuming A_n we know that $K_X \otimes L^{n-1}$ is semi-ample.

Choose M large enough and divisible by r so that $\Gamma((K_X \otimes L^{n-1})^M)$ spans $(K_X \otimes L^{n-1})^M$ and so that the holomorphic map $\varphi : X \longrightarrow \mathbb{P}$ associated to the space of sections has connected fibres and normal image $Y = \varphi(X)$.

If $\dim \varphi(X) = 0$ then $(K_X \otimes L^{n-1})^M \cong \mathcal{O}_X$. Choose a general $A \in |L|$ such that A is normal and $(K_X \otimes L)_A \cong K_A$. We see that $(K_A \otimes L_A{}^{n-2})^M \cong \mathcal{O}_A$. Looking over the table for $\dim A = n-1$ we see that $K_A \otimes L_A{}^{n-2} \cong \mathcal{O}_A$. Therefore by (0.3.1), $K_X \otimes L^{n-1} \cong \mathcal{O}_X$.

Let $\dim \varphi(X) = 1$. Choose a general $A \in |L|$ such that A is normal and $(K_X \otimes L)_A \cong K_A$. By the table for $\dim A = n-1$ we see that $K_A \otimes L_A{}^{n-2} \cong \varphi^*\mathfrak{L}$ for an ample line bundle $\mathfrak{L}$ on Y. Therefore $(K_X \otimes L^{n-1}) \otimes (\varphi^*\mathfrak{L})^{-1})_A \cong \mathcal{O}_A$. By (2.2.1), $K_X \otimes L^{n-1} \cong \varphi^*\mathfrak{L}$ and (X,L) is a scroll.

Let $\dim \varphi(X) = 2$. By the same reasoning as before when $\dim \varphi(X) = 1$, we see that $(K_X \otimes L^{n-1})^r \cong \varphi^*\mathfrak{L}$ for an ample line bundle $Y = \varphi(X)$ where r is the index of Y. (X,L) is a quadric bundle by definition.

Assume that $\dim \varphi(X) \geqslant 3$ and $\neq n$. Let (F,L_F) be a generic fibre of φ. As before F is a normal variety and satisfies $(K_F \otimes L_F{}^{\dim F + 2})^r \cong \mathcal{O}_F$. Since $\mathrm{Irr}(X)$ is finite, it follows that F can be chosen with $\mathrm{Irr}(F)$ empty. Thus we have a contradiction to (1.2) applied to (F,L_F).

□

There are a number of remarkable corollaries of the above result. The first follows immediately from Sakai's results for $n = 2$ (see [Sa3]).

(4.4.3) **Corollary**. Let (X,L) be as in (4.0). If

$$(K_X + (n-1)L)\cdot(K_X + (n-1)L)\cdot \underbrace{L \cdot \;\cdot\;\cdot\; \cdot L}_{n-2 \text{ copies}} \geqslant 5$$

then $K_X \otimes L^{n-1}$ is nef and big.

(4.4.4) **Corollary**. Let (X,L) be as in (4.0). If $\mathrm{Irr}(X)$ is non-empty then either (X,L) is a Gorenstein Del Pezzo surface with one elliptic singularity or $K_X \otimes L^{n-1}$ nef and big.

Proof. Inspection of the table in (4.4).

□

(4.5) **Theorem**. Let (X,L) be as in (4.0). Assume further that X is Gorenstein, $\mathrm{cod}\,\mathrm{Sing}(X) \geqslant 3$, and $n \geqslant 2$. Assume that $K_X \otimes L^{n-1}$ is nef and big. Then there is a reduction (X',L') of

(X,L) <u>such that</u> $K_{X'} \otimes L'^{n-1}$ <u>is ample. If</u> $h^0((K_X \otimes L^{n-2})^N) \neq 0$ <u>for some</u> $N > 0$, <u>then</u> $K_{X'} \otimes L'^{n-2}$ <u>is nef and any smooth surface</u> S <u>which is the intersection of</u> n-2 <u>general members of</u> |L'| <u>is a minimal model of non-negative Kodaira dimension.</u>

Proof. The above result was proven for $n = 3$ in [F+S] (see also [So6]) when $h^0((K_X \otimes L)^N) \neq 0$ for $N > 0$. Knowing that $K_X \otimes L^{n-1}$ is semi-ample and that L is spanned by global sections it is straightforward to deduce (3.2) from the arguments in ([F+S], [L+S]) and the result (2.1) on the ascent of numerical effectivity.

□

(4.5.1) **Remark**. If we merely assume that X is Cohen-Macaulay and $\mathbb{Q}$-Gorenstein, with Irr(X) and the set of non-Gorenstein points finite, and with cod Sing(X) ≥ 3, then most of the above is still true. The exception is that I can only show that $K_{X'} + (n-1 - 1/r)L'$ is nef where r is the index of X. The key point in this generalization is the observation to me by F.Sakai that theorem (0.1) of [L+S] holds for normal $\mathbb{Q}$-Gorenstein surfaces [Sa3].

(4.5.2) **Corollary**. <u>Let</u> (X,L) <u>and</u> (X',L') <u>be as in</u> (4.5). <u>If</u> $K_{X'} \otimes L'^{n-2}$ <u>is big and nef or if the Kodaira dimension of the surface</u> S <u>in</u> (4.5) <u>is</u> 0 <u>then</u> $K_{X'} \otimes L'^{n-2}$ <u>semi-ample.</u>

Proof. By using (2.1) and (2.2) we can assume that $n = 3$; the case when $n = 2$ is very classical. Consider :

$$0 \longrightarrow K_{X'} \otimes (K_{X'} \otimes L')^N \longrightarrow (K_{X'} \otimes L')^{N+1} \longrightarrow K_S^{N+1} \longrightarrow 0$$

where $S \in |L|$ is smooth and K_S^N is spanned for $N >> 0$. Since $(K_{X'} \otimes L')$ is nef and big we conclude by (0.2.1b) that $\Gamma((K_{X'} \otimes L')^{N+1})$ spans K_S^{N+1}. By the Fujita-Zariski theorem [F5], $K_{X'} \otimes L'$ is semi-ample.

If $K_S^N \cong \mathcal{O}_S$ then either $\Gamma(K_{X'} \otimes L') \neq 0$ or $\Gamma(K_{X'} \otimes L') = 0$. In the former case we conclude by (0.3.1) that $K_{X'} \otimes L' \cong \mathcal{O}_{X'}$. In the latter case we know by (0.2.2) that X' has rational singularities. Rational Gorenstein singularities are canonical [Ka3] and therefore $K_{X'} \otimes L'$ is semi-ample by [Ka3].

□

§5 The Spectral Classes $\mathcal{K}_\sigma$ for $\sigma < 3$

(5.0) Throughout this section L will be an ample and spanned line bundle on X, a normal, Gorenstein projective variety of dimension $n \geq 2$ with Irr(X) finite and cod Sing(X) ≥ 3. Assume also that Irr(X) is empty if dim X = 3.

(5.1) **Theorem**. Let (X,L) be as above. Assume that $K_X \otimes L^{n-1}$ is nef and big and let (X',L') be the reduction of (4.5). Then the following are equivalent:

a) $h^0((K_X \otimes L^{n-2})^N) \neq 0$ for some $N > 0$,

b) $K_{X'} \otimes L'^{n-2}$ is nef,

c) $K_{X'} \otimes L'^{n-2}$ is semi-ample

d) there is a smooth surface S which is the transverse intersection of n-2 generic elements of $|L'|$ such that S is a minimal model of non-negative Kodaira dimension.

Proof. Since the above is very classical if $n = 2$, we can assume without loss of generality that $n \geq 3$. As pointed out earlier in (4.5) a) implies b) and a) implies d). It is easy to check that b) implies d). Clearly c) implies a). Therefore we need only show that b) implies c) and d) implies b).

Assume b). By using (2.1) and (2.2) we can assume that $n = 3$. As in (4.5.2) this reduces to [Ka3].

Assume d) holds but not b). By (2.1) we can assume that $n = 3$. If $(K_{X'} \otimes L')^t \otimes L'^N$ is semi-ample for all $t \geq 0$ where N is fixed it follows that $K_{X'} \otimes L'$ is nef. Indeed given an effective curve C:

$$(K_{X'} + L')\cdot C = \lim_{t \to \infty} [(t(K_{X'} + L') + NL')\cdot C]/t \geq 0.$$

Thus for any N there is a greatest integer t such that $\mathcal{L}(t) = (K_{X'} \otimes L')^t \otimes L'^N$ is semi-ample. I claim that there is an $N > 0$ so that $\mathcal{L}(t)$ isn't big for the last t for which $\mathcal{L}(t)$ is semi-ample. Let $S \in |L'|$. By (0.6) it follows that given any ample line bundle on a minimal model surface of non-negative Kodaira dimension, say L'_S on S, there is an $N > 0$ such that $K_S{}^t \otimes L'_S{}^N$ is ample and spanned for all $t \geq 0$. Consider:

$$*_t) \quad 0 \longrightarrow K_{X'} \otimes (K_{X'} \otimes L')^t \otimes L'^N \longrightarrow (K_{X'} \otimes L')^{t+1} \otimes L'^N \longrightarrow K_S{}^{t+1} \otimes L'_S{}^N \longrightarrow 0.$$

If $\mathcal{L}(t)$ is nef and big then by the above, (0.2.1b), and the Fujita-Zariski theorem it follows that $\mathcal{L}(t+1)$ is semi-ample. Therefore we have a pair of positive integers $a < b$ such that:

a) $\Gamma(K_{X'}{}^a \otimes L'^b)$ spans $K_{X'}{}^a \otimes L'^b$,

b) the map φ associated to $\Gamma(K_{X'}{}^a \otimes L'^b)$ has an image of dimension $< n = 3$,

c) φ has connected fibres and normal image $Y = \varphi(X')$.

If this happens we have by the usual reasoning an irreducible and smooth fibre F of positive dimension such that $K_F{}^a \otimes L'_F{}^b \cong \mathcal{O}_F$. Since $K_{X'} \otimes L'^2$ is ample we know that $b < 2a$. We can assume by (0.3.1) that a and b are relatively prime. Therefore by writing $qa+pb = 1$ for integers p, q we see that there is an ample line bundle $M = L'_F{}^q \otimes K_F{}^{-p}$ so that $M^b = K_F{}^{-1}$ and $M^a = L'_F$. Since $2a > b > a > 0$ we see that $\dim F \neq 1$ and if $\dim F = 2$ then $b = 3$ and $F = \mathbb{P}^2$, $L'_F \cong \mathcal{O}_{\mathbb{P}^2}(2)$. But in this case S is birationally ruled which is absurd. Therefore $\dim F = 3$, and $b = 3$ or 4. In the first case X is a quadric and K_S is negative. In the second case $X = \mathbb{P}^3$ and K_S is negative.

□

The following can be deduced by reasoning of exactly the same sort as used in § 3.

(5.2) **Theorem**. Let (X,L) be as in (5.0). Assume that $h^0((K_X \otimes L^{n-2})^N) \neq 0$ for some $N > 0$ and let (X',L') be the reduction of (4.5). Let $\log_3$ (X,L) be the dimension of the image of X under the map $\varphi_2 : X \longrightarrow Y$ associated to $\Gamma((K_{X'} \otimes L'^{n-2})^N)$ for large N. There is the following table. Assume that $n \geq 3$.

$\log_3$(X,L)	Structure of (X,L)
0	$K_X \cong L'^{-(n-2)}$
1	Del Pezzo bundle over a curve
2	Quadric bundle over a surface
$3 \neq n$	Scroll over a 3 fold
> 3 and < n	does not occur

The following theorem combined with the results of § 4 gives very good information about the spectral class $\mathfrak{X}_3$.

(5.3) **Theorem**. Let L be an ample spanned line bundle on a normal, Gorenstein variety with Irr(X) finite. $(X,L) \in \mathfrak{X}_3$ if and only if

a) $K_X \otimes L^{n-1}$ is not nef and big,

or

b) X is a birationally ruled surface

or

c) there is a holomorphic surjective map $\varphi : X' \longrightarrow C$ with connected fibres onto a smooth curve, C, and such that the general fibre of φ is $(\mathbb{P}^2, \mathcal{O}_{\mathbb{P}^2}(2))$ with $K_{X'}^2 \otimes L'^3 \cong \varphi^* \mathcal{S}$ for some ample line bundle $\mathcal{S}$ on C,

or

d) $(X',L') = (\mathbb{Q}, \mathcal{O}_{\mathbb{Q}}(2))$ where $(\mathbb{Q}, \mathcal{O}_{\mathbb{Q}}(1))$ is a 3 dimensional quadric,

or

e) $(X',L') = (\mathbb{P}^3, \mathcal{O}_{\mathbb{P}^3}(3))$,

or

f) $(X',L') = (\mathbb{P}^4, \mathcal{O}_{\mathbb{P}^4}(2))$,

where in c) through f) $K_X \otimes L^{n-1}$ is nef and big and (X',L') is the reduction associated to $K_X \otimes L^{n-1}$.

Proof. If (X,L) is as in a) through f) then $h^0((K_X \otimes L^{n-2})^N) = 0$ for all $N > 0$ by (5.1) and inspection. Therefore it can be assumed without loss of generality that $h^0((K_X \otimes L^{n-2})^N) = 0$ for all $N > 0$. We can assume that $n \geq 3$ since otherwise the result is classical. By (0.2.2) Irr(X) is empty. Also it can be assumed without loss of generality that $K_X \otimes L^{n-1}$ is nef and big. Let (X',L') be the reduction associated to (X,L) by $K_{X'} \otimes L'^{n-1}$. Since $K_X \otimes L^{n-1}$ is nef and big we know that $K_{X'} \otimes L'^{n-1}$ is ample.

Assume first that $n = 3$. For smooth threefolds the desired result was proved in [So6]; here we give essentially the same proof.

Note that (X',L') is one of the examples in c) through e) above if we have a surjective mapping $\varphi : X' \longrightarrow Y$ with $K_{X'}^a \otimes L'^b \cong \varphi^* \mathcal{L}$ where :

1) φ has connected fibres of positive dimension and $\mathcal{L}$ is an ample line bundle on a normal variety Y,

2) $0 < a < b < 2a$.

Therefore to show the above theorem when $n = 3$ it suffices to show that $h^0((K_X \otimes L^{n-2})^N) = 0$ for all $N > 0$ implies that there are $0 < a < b < 2a$ with $K_{X'}^a \otimes L'^b$ semi-ample and not big.

Let $S \in |L'|$ be a smooth surface. By the fundamental theorem (0.6) we can choose an $N > 0$ such that $\Gamma(K_S^t \otimes L'_S{}^N)$ spans $K_S^t \otimes L'_S{}^N$ for $0 \leq t \leq A$ where either :

α) $K_S^A \otimes L'_S{}^N$ is not ample,

or

β) $2A = N$,

or

γ) $K_S^t \otimes L'_S{}^N \cong \mathcal{O}_S$ for some $a > 0$.

Since $K_{X'} \otimes L'^2$ is ample, β) is ruled out. Case γ) implies that (X',L') is as in (5.3d) or (5.3e). Indeed by (0.3.1) $K_{X'}{}^a \otimes L'^{N+a} \cong \mathcal{O}_{X'}$. Therefore after dividing out by common factors $K_{X'}{}^a \otimes L'^b \cong \mathcal{O}_{X'}$ for $0 < a < b < 2a$. Therefore we can assume that $K_S{}^t \otimes L'_S{}^N$ is spanned for $0 \leqslant t \leqslant A$ with $K_S{}^A \otimes L'_S{}^N$ not ample.

By the same argument as in (5.1) with the sequences $*_t$) we arrive at some $t > 0$ and $\leqslant A$ such that either :

a) $\mathcal{L}(t)$ is semi-ample but not big,

or

b) $\mathcal{L}(t)$ is semi-ample and big but $\mathcal{L}(t)_S \cong K_S{}^t \otimes L'_S{}^N$ is not ample.

For the given t let $\varphi : X' \longrightarrow Y$ be the map associated to $\Gamma((K_{X'}{}^t \otimes L'^{t+N})^{\mathbb{N}})$ for $\mathbb{N}$ large enough so that Y is normal and φ has connected fibres.

In case a) let F be a generic positive dimensional fibre of φ. We have $K_F{}^a \otimes L'_F{}^b \cong \mathcal{O}_F$ with $0 < a < b < 2a$ and a, b relatively prime. By the argument of (5.1) there exists an ample line bundle M such that $M^b \cong K_F{}^{-1}$ and $M^a \cong L'_F$. From this it is easy to conclude that (X',L') is in either (5.3c), (5.3d), or (5.3e).

In case b) we know that $\varphi_S : S \longrightarrow Y$ has some positive dimensional fibres. If $\dim \varphi(S) < \dim S$ then by (0.3.2) $\dim \varphi(X) = \dim \varphi(S)$. In this case $\mathcal{L}(t)$ is not big. Using (0.2.1b) and the fact that $\mathcal{L}(t)$ is big we see that $H^1(K_{X'} \otimes \mathcal{L}(t)^{\mathbb{N}}) = 0$ for $\mathbb{N} > 0$ and therefore by $*_t$) that for all $\mathbb{N} > 0$ $\Gamma(\mathcal{L}(t)^{\mathbb{N}}) \longrightarrow \Gamma(\mathcal{L}(t)_S{}^{\mathbb{N}}) \longrightarrow 0$.

Therefore φ_S is the map associated to $\Gamma(K_S{}^{(t+N)\mathbb{N}} \otimes L'_S{}^{N\mathbb{N}})$ for large $\mathbb{N}$. Therefore by [L+F], [So8], or [Sa3] we can see that $\varphi(S)$ is smooth and S is the blowup of $\varphi(S)$ at a finite set of points F. Let $E \subset S$ be such that $\varphi(E) \in F$. We know that :

$$**) \qquad E \cdot E = -1, \qquad K_S \cdot E = -1, \qquad ((t+N-1)K_S + NL'_S) \cdot E = 1.$$

From the deformation theory argument of [So3] we see that the closure of the union of the deformations of E in X' is either X' or a divisor. Since $\dim \varphi(X') = 3$ we get,a divisor $\mathcal{D}$. We know that it is normal with S meeting $\mathcal{D}$ transversely in E by the argument of [So3]. Therefore $\mathcal{D}$ is either F_r or F_r in the notation of [So3]. Note that $L'_{\mathcal{D}} = [E]$, and $\mathcal{E} = (K_S{}^{t+N-1} \otimes L'_S{}^N)_{\mathcal{D}}$ is ample and by $**$) it has degree 1 on E. Therefore $L'_{\mathcal{D}} \cdot \mathcal{E} = 1$ and by an easy calculation (as in [So3]) we see that $(\mathcal{D}, L'_{\mathcal{D}}) \cong (\mathbb{P}^2, \mathcal{O}_{\mathbb{P}^2}(1))$. But then $K_{X'} \otimes L'^2)_{\mathcal{D}} \cong \mathcal{O}_{\mathcal{D}}$, i.e. $K_{X'} \otimes L'^2$ is not ample. By this contradiciton we see that case a) doesn't happen.

Assume now that $n \geqslant 4$ and $h^0((K_X \otimes L^{n-2})^N) = 0$ for all $N > 0$. Choose a general $A \in$

$|L|$. If $h^0((K_A \otimes L_A{}^{n-3})^N) = 0$ for all $N > 0$ then by induction we can assume that (A,L_A) has a reduction $(A',L_{A'})$ as in (5.3c) through (5.3f). This assertion only has meaning for $n = 4$ or 5. Let us consider these cases.

Assume that $n = 4$. Assume that $(A',L_{A'})$ is as in (5.3c). Then $K_{A'}{}^2 \otimes L_{A'}{}^3$ is semi-ample and therefore by (2.1) $K_{X'}{}^2 \otimes L'^5$ is nef. From $\Gamma(K_X \otimes L) = 0$ we conclude by (0.2.2) that X has only rational singularities. Since rational Gorenstein singularities are canonical we conclude by [Ka3] that $K_{X'}{}^2 \otimes L'^5$ is semi-ample. The map ψ associated to $\Gamma((K_{X'}{}^2 \otimes L'^5)^N)$ for $N >> 0$ extends $\varphi : A \longrightarrow C$ with the same curve as image, e.g. use the argument of (2.2.1). But on a general fibre F of ψ, $K_F{}^2 \otimes L'_F{}^5 \cong \mathcal{O}_F$ which implies that $K_F{}^{-1} = M^5$ for some ample line bundle. This contradicts (1.2) since dim F = 4. In case $(A',L'_{A'})$ is as in d); we get by (0.3.1) that $K_{X'}{}^2 \otimes L'^5 \cong \mathcal{O}_{X'}$. Again we have an ample line bundle M such that $M^5 \cong K_X{}^{-1}$. We conclude that (X',L') is as in f).

If $(A',L'_{A'})$ is as in e) then we get that there is an ample line bundle M on X' such that $M^7 = K_{X'}{}^{-1}$. By (1.2) this can't happen.

If $n = 5$ then only f) must be considered as a possibility for $(A',L'_{A'})$. In this case as before $K_{X'}{}^2 \otimes L'^7 \cong \mathcal{O}_{X'}$ and there is an ample line bundle M such that $M^7 = K_{X'}{}^{-1}$. This can't happen by (1.2).

□

The above results have many applications. Here we restrict ourselves to a single example.

(5.4) **Theorem**. Let S be a smooth Kodaira dimension zero surface. If $S \in |L|$ for some ample spanned line bundle L on a 3 dimensional normal Gorenstein projective variety with cod Sing(X) $\geqslant 3$ then there is a reduction (X',L') of (X,L) such that :

a) S is a K-3 surface with $K_{X'} \cong L'^{-1}$,

or

b) (X,L) is a scroll over a surface birational to S.

Proof. If $h^0((K_X \otimes L)^N) = 0$ for some $N > 0$ then by inspection of the elements of the class $\mathfrak{X}_3$ we see that (X,L) is as in b). If $h^0((K_X \otimes L)^N) > 0$ for some $N > 0$, then by (4.5) there is a reduction (X',L') of (X,L) with $K_{X'} \otimes L'$ nef. Since $(K_{X'} \otimes L')_{S'}{}^N \cong \mathcal{O}_{S'}$ where S' is associated to S by the reduction we conclude that $(K_{X'} \otimes L')^N \cong \mathcal{O}_{X'}$ by (0.3.1). We claim that we can choose $N = 1$. Assume otherwise that $N > 1$ is the smallest integer > 0 such that $(K_{X'} \otimes L')^N \cong \mathcal{O}_{X'}$. Let $\alpha : Y \longrightarrow X$ be the non-trivial N sheeted unbranched cover associated to the Nth root of the trivializing section of $(K_{X'} \otimes L')^N$. Note that $\chi(Y,\mathcal{O}_Y) = N\chi(X',\mathcal{O}_{X'})$. Since $K_{X'} \cong L'^{-1} \otimes (K_{X'} \otimes L')^{-N+1}$ we see that $\chi(X',\mathcal{O}_{X'}) - h^0(\mathcal{O}_{X'})$ by (0.2.1b). Similarly $\chi(Y,\mathcal{O}_Y) = h^0(\mathcal{O}_Y) = 1$. This implies that $N = 1$. (It is my understanding that this sort of argument was used first by S. Kobayashi). Note that we have also concluded that $h^1(\mathcal{O}_{X'}) = h^2(\mathcal{O}_{X'}) = 0$. From this $K_{X'} \cong L'^{-1}$, (0.2.1b), and the residue sequence

$$0 \longrightarrow K_{X'} \longrightarrow \mathcal{O}_{X'} \longrightarrow \mathcal{O}_{S'} \longrightarrow 0$$

we conclude that $h^1(\mathcal{O}_{S'}) = 0$, $h^2(\mathcal{O}_{S'}) = 1$. From this we conclude that S' is a K-3 surface. □

In the same vein one can analyze (X,L) with $S \in |L|$ an elliptic surface. We did this in [So6] for X smooth and $h^1(\mathcal{O}_X) > 0$. Using the assumption that Irr(X) is empty and using theorem (5.1) in place of the Albanese mapping, the proof in [So6] yields that if $h^0((K_X \otimes L)^N) \neq 0$ for some $N > 0$ then the elliptic fibration $S \longrightarrow C$ has **no multiple fibres**. Since the proof of this would take us far afield, I will not include it here.

References

[A+K] A. Altman and S. Kleiman, Introduction to Grothendieck Duality Theory, Springer Lecture Notes 146 (1970), Springer Verlag, Berlin.

[B+P] M. Beltrametti and M. Palleschi, On threefolds with low sectional genus, preprint.

[(F+S)1] M.L. Fania and A.J. Sommese, On the minimality of hyperplane sections of Gorenstein 3-folds, to appear Proceedings in honor of W. Stoll, Notre Dame (1984), Vieweg.

[(F+S)2] M.L. Fania and A.J. Sommese, Varieties whose hyperplane sections are $\mathbb{P}^k$ bundles, preprint.

[F1] T. Fujita, On the structure of polarized varieties of Δ genera zero, J. Fac. Sci. Univ. of Tokyo, 22, 103-115 (1975).

[F2] T. Fujita, On the hyperplane section principle of Lefschetz, J. Math. Soc. Japan 32 (1980), 153-169.

[F3] T. Fujita, On the structure of Polarized manifolds of total deficiency one, J. Math. Soc. Japan 32 (1980), 709-725.

[F4] T. Fujita, Rational retractions onto ample divisors, Sci. Papers of the College of

Arts and Sciences of the University of Tokyo 33 (1983), 33-39.

[F5] T. Fujita, Semi-positive line bundles, Jour. Univ. Tokyo, Sect. IA Math. 30 (1983), 353-378.

[F6] T. Fujita, Generalized Adjunction Mappings, preprint of research notice.

[F7] T. Fujita, Projective varieties of Δ-genus 1, preprint.

[G+R] H. Grauert and O. Riemenschneider, Verschwindungssatze fur analytische Kohomologiegruppen auf komplexen Raumen, Inv. Math. 11 (1970), 263-292.

[Io] P. Ionescu, On varieties whose degree is small with respect to codimension, Math. Ann. 271 (1985), 339-348.

[Ka1] Y. Kawamata, A generalization of Kodaira-Ramanujam's vanishing theorem, Math. Ann. 261 (1982), 43-46.

[Ka2] Y. Kawamata, Elementary contractions of algebraic 3-folds, Ann. of Math. 119 (1984), 95-110.

[Ka3] Y. Kawamata, The cone of curves of algebraic varieties, Ann. of Math. 119 (1984), 603-633.

[Ke] G. Kempf et al., Toroidal embeddings I, Springer Lecture Notes 339 (1973), Springer Verlag, Berlin.

[L+P] A. Lanteri and M. Palleschi, About the adjunction process for polarized algebraic surfaces, J. reine und angew. Math. 352 (1984), 15-23.

[L+S] J. Lipman and A.J. Sommese, On the contraction of projective spaces on singular varieties, to appear J. reine und angew. Math.

[M] S. Mori, Threefolds whose canonical bundles are not numerically effective, Ann. of Math. 116 (1982), 133-176.

[Re] M. Reid, Canonical 3-folds, Geometric Algebrique d'Angiers (ed. A. Beauville), 273-310, Alphen aan den Rijn Sijthoof and Noordhoff (1980).

[Ro] L. Roth, Algebraic Threefolds, Springer-Verlag, Berlin (1953).

[Sa1] F. Sakai, Semi-stable curves on algebraic surfaces and logarithmic pluricanonical maps, Math. Ann. 254, 89-120 (1980).

[Sa2] F. Sakai, The structure of normal surfaces, preprint.

[Sa3] F. Sakai, Ample Cartier divisors on normal surfaces, preprint.

[Sh+So] B. Shiffman and A.J. Sommese. Vanishing Theorems on Complex Manifolds, Progress in Mathematics 56 (1985), Birkhauser, Boston.

[So1] A.J. Sommese, On manifolds that cannot be ample divisors, Math. Ann. 221 (1976), 55-72.

[So2] A.J. Sommese, Hyperplane sections of projective surfaces I- The Adjunction Mapping, Duke Math. J., 46 (1979), 377-401.

[So3] A.J. Sommese, On the minimality of hyperplane sections of projective threefolds, J. reine und angew. Math. 329 (1981), 16-41.

[So4] A. J. Sommese, Hyperplane sections, Algebraic Geometry, Proceedings Chicago Circle Conference, 1980, Springer Lecture Notes 862 (1981), 232-271.

[So5] A.J. Sommese, Ample divisors on 3-folds, Algebraic Threefolds, Springer Lecture Notes 947 (1982), 229-240.

[So6] A.J. Sommese, On the birational theory of hyperplane sections of projective threefolds, Unpublished 1981 Manuscript.

[So7] A.J. Sommese, Configurations of -2 rational curves on hyperplane sections of projective threefolds, Classification of Algebraic and Analytic Manifolds (ed. K. Ueno), Progress in Mathematics (1983), Birkhauser, Boston.

[So8] A.J. Sommese, Ample divisors on normal Gorenstein surfaces, to appear Abh. Math. Hamburg.

[So9] A.J. Sommese, Ample divisors on Gorenstein varieties, to appear Proceedings of Complex Geometry Conference, Nancy 1985.

[V] E. Viehweg, Vanishing theorems, J. reine und angew. Math. 335 (1982), 1-8.

Value Distribution Theory for Moving Targets

Wilhelm Stoll*

About sixty years ago, Nevanlinna [9] created value distribution theory in extension of the theory of entire functions. Here some progress in the value distribution theory of several complex variables for moving targets shall be reported. The theory works on parabolic manifolds, but for simplicity $\mathbb{C}^m$ alone shall be considered as domain space.

First the theory for fixed targets shall be outlined. The **norm** of $\mathfrak{z} = (z_1, \ldots, z_m) \in \mathbb{C}^m$ is denoted by

(1) $$\|\mathfrak{z}\| = (|z_1|^2 + \ldots + |z_m|^2)^{1/2} .$$

If $r > 0$ and $S \subseteq \mathbb{C}^m$ abbreviate

(2) $$S(r) = \{\mathfrak{z} \in S \mid \|\mathfrak{z}\| < r\}$$

(3) $$S[r] = \{\mathfrak{z} \in S \mid \|\mathfrak{z}\| \leq r\}$$

(4) $$S\langle r\rangle = \{\mathfrak{z} \in S \mid \|\mathfrak{z}\| = r\} .$$

Define $\tau : \mathbb{C}^m \longrightarrow \mathbb{R}$ by $\tau(\mathfrak{z}) = \|\mathfrak{z}\|^2$ for all $\mathfrak{z} \in \mathbb{C}^m$. Normalize $d^c = (i/4\pi)(\bar{\partial} - \partial)$. Define

(5) $$\upsilon = dd^c\tau \qquad \omega = dd^c \log \tau$$

(6) $$\sigma = d^c \log \tau \wedge \omega^{m-1} .$$

Then $\upsilon > 0$ and $d\sigma = 0$. Thus $(\mathbb{C}^m, \tau)$ is a strictly parabolic manifold with

*This research was supported in part by the National Science Foundation NSF Grant DMS 84-04921.

(7) $$\int_{\mathbb{C}^m(r)} \upsilon^m = r^{2m} \int_{\mathbb{C}^m\langle r\rangle} \sigma = 1 .$$

A **divisor** on $\mathbb{C}^m$ can be defined as an integral valued function $\nu : \mathbb{C}^m \longrightarrow \mathbb{Z}$ whose support $S = \operatorname{supp} \nu$ is either empty or an analytic set of pure dimension $m - 1$. For $r > 0$ the **counting function** n_ν of ν is defined by

(8) $$n_\nu(r) = r^{2-2m} \int_{S[r]} \nu\upsilon^{m-1} = \int_{S[r]} \nu\omega^{m-1} + \nu(0) .$$

If $\nu \geqslant 0$, the counting function is non-negative and increases. Moreover

(9) $$n_\nu(\infty) = \lim_{\nu\to\infty} n_\nu(r)$$

exists. Here $n_\nu(\infty) < \infty$ if and only if n_ν is affine algebraic. For $0 < s < r$, the **valence function** N_ν of any divisor $\nu : \mathbb{C}^m \longrightarrow \mathbb{Z}$ is defined by

(10) $$N_\nu(r,s) = \int_s^r n_\nu(t) \frac{dt}{t} .$$

The divisor of the zeros of an entire function $h \not\equiv 0$ is denoted by $\mu_{h,0}$.

Let V be a hermitian vector space of finite dimension $n + 1$. For $\mathfrak{z} \in V$ and $\mathfrak{w} \in V$, the hermitian scalar product of $\mathfrak{z}$ and $\mathfrak{w}$ is denoted by $(\mathfrak{z} \mid \mathfrak{w})$ and the norm of $\mathfrak{z}$ by $\|\mathfrak{z}\| = (\mathfrak{z} \mid \mathfrak{z})^{1/2}$. Put $V_* = V - \{0\}$. Then $\mathbb{P}(V) = V_*/\mathbb{C}_*$ is the **projective space** associated to V. Let $\mathbb{P} : V_* \longrightarrow \mathbb{P}(V)$ be the residual map. For $S \subseteq V$, define $\mathbb{P}(S) := \mathbb{P}(S - \{0\})$. The hermitian scalar product on V determines uniquely a **Fubini-Study form** Ω on $\mathbb{P}(V)$ with

(11) $$\int_{\mathbb{P}(V)} \Omega^n = 1 .$$

The dual vector space V^* of V consists of all $\mathbb{C}$-linear maps $\alpha : V \longrightarrow \mathbb{C}$ and inherits a hermitian scalar product from V. If $\mathfrak{x} \in V$ and $\alpha \in V^*$ their **inner product** is also denoted by

(12) $$\langle \mathfrak{z},\alpha\rangle = \alpha(\mathfrak{z}) .$$

For $a \in \mathbb{P}(V^*)$, take $\alpha \in V^*_*$ with $\mathbb{P}(\alpha) = a$. Then $E[a] = \mathbb{P}(\ker \alpha)$ is a hyperplane in $\mathbb{P}(V)$ and $\mathbb{P}(V^*)$ bijectively parameterizes all hyperplanes in $\mathbb{P}(V)$. If $x = \mathbb{P}(\mathfrak{z}) \in \mathbb{P}(V)$, the distance from x to $E[a]$ is given by

(13) $$0 \leqslant \Box\, x;a\, \Box = \frac{|\langle \mathfrak{z},\alpha\rangle|}{\|\mathfrak{z}\|\ \|\alpha\|} \leqslant 1 .$$

Let $f : \mathbb{C}^m \longrightarrow \mathbb{P}(V)$ be a meromorphic map. A holomorphic map $\mathfrak{w} : \mathbb{C}^m \longrightarrow V$ is said to be a **representation** of f if $\mathfrak{w} \not\equiv 0$ and if $f = \mathbb{P} \circ \mathfrak{w}$ on $\mathbb{C}^m - \mathfrak{w}^{-1}(0)$. The representation is said to be **reduced** if $\dim \mathfrak{w}^{-1}(0) \leqslant m - 2$. If $\mathfrak{w}$ is a representation of f, then there is a reduced representation $\tilde{\mathfrak{w}}$ of f and an entire function h such that $\mathfrak{w} = h\tilde{\mathfrak{w}}$. The non-negative divisor $\mu_{\mathfrak{w}} = \mu_{h,0}$ does not depend on the choice of $\tilde{\mathfrak{w}}$ and h. A meromorphic map $f : M \longrightarrow \mathbb{P}(V)$ admits a reduced representation $\mathfrak{w} : \mathbb{C}^m \longrightarrow V$. Let $\mathfrak{w}$ be a reduced representation of f. Then there are holomorphic vector functions $\mathfrak{w}_j : \mathbb{C}^m \longrightarrow V$ such that $\mathfrak{w}_j(a\mathfrak{z}) = a^j \mathfrak{w}(\mathfrak{z})$ for all $\mathfrak{z} \in \mathbb{C}^m$ and $a \in \mathbb{C}$, such that $\mathfrak{w}_p \not\equiv 0$ and such that $\mathfrak{w} = \sum_{j=p}^{\infty} \mathfrak{w}_j$ converges uniformly on every compact subset of $\mathbb{C}^m$. The non-negative integer p does not depend on the choice of $\mathfrak{w}$ and is denoted by $A_f(0)$.

For $r > 0$, the **spherical image function** A_f is defined by

(14) $$A_f(r) = r^{2-2m} \int_{\mathbb{C}^m[r]} f^*(\Omega) \wedge \upsilon^{m-1} .$$

The function A_f in non-negative and increases with r. Then

(15) $$A_f(r) = \int_{\mathbb{C}^m[r]} f^*(\Omega) \wedge \omega^{m-1} + A_f(0) .$$

Define

(16) $$A_f(\infty) = \lim_{r\to\infty} A_f(r) .$$

Then $A_f(\infty) = 0$ if and only if f is constant. Also $A_f(\infty) < \infty$ if and only if f is rational. For $0 < s < r$ the **characteristic function** T_f of f is defined by

$$T_f(r,s) = \int_s^r A_f(t)\,\frac{dt}{t} \tag{17}$$

where $T_f(r,s)/\log r \longrightarrow A_f(\infty)$ for $r \longrightarrow \infty$. Hence $T_f(r,s) \longrightarrow \infty$ for $r \longrightarrow \infty$ if and only if f is not constant.

Take $a \in \mathbb{P}(V^*)$. Then (f,a) is said to be **free** if $f(\mathbb{C}^m) \not\subseteq E[a]$, which is the case if and only if $\square\, f,a\, \square \not\equiv 0$. If so, the **compensation function** $m_{f,a}$ is defined for $r > 0$ by

$$m_{f,a}(r) = \int_{\mathbb{C}^m\langle r\rangle} \log \frac{1}{\square\, f,a\, \square}\,\sigma \geqslant 0\,. \tag{18}$$

Take $\alpha \in V_*^*$ with $\mathbb{P}(\alpha) = a$. Let $\mathfrak{v}$ be a reduced representation of f. Then (f,a) is free if and only if $\langle \mathfrak{v}, \alpha\rangle \not\equiv 0$. If so, the **intersection divisor** $\mu_{f,a} = \mu_{\langle \mathfrak{v},\alpha\rangle} \geqslant 0$ does not depend on the choices of α and $\mathfrak{v}$. Its counting function is denoted by $n_{f,a}$ and its valence function is denoted by $N_{f,a}$.

Assume that (f,a) is free. Take real numbers r and s with $0 < s < r$. The **First Main Theorem** (FMTH) states

$$\boxed{T_f(r,s) = N_{f,a}(r,s) + m_{f,a}(r) - m_{f,a}(s)} \tag{19}$$

Thus the growth measure of the intersection divisor of f with E[a] plus the compensation terms balance with the growth measure of the map f. The relative distribution of the contributions for large r is measured by the **defect**

$$0 \leqslant \delta(f,a) = \lim_{r\to\infty} \frac{m_{f,a}(r)}{T_f(r,s)} = 1 - \lim_{r\to\infty} \frac{N_{f,a}(r,s)}{T_f(r,s)} \leqslant 1 \tag{20}$$

provided f is not constant. Thus $\delta(f,a) = 1$, if $f(\mathbb{C}^m) \cap E[a] = \emptyset$, or if $\mu_{f,a}$ is affine algebraic and f is transcendental.

The map f is said to be **linearly non-degenerate**, if $f(\mathbb{C}^m)$ is not contained in any hyperplane. A subset $\mathfrak{g}$ of $\mathbb{P}(V^*)$ is said to be in **general position**, if any subset $\mathfrak{h}$ of $\mathfrak{g}$ with $\#\mathfrak{h} \leqslant n + 1$ is linearly independent. If both is assumed, the **Defect Relation** holds

(21)

$$\boxed{\sum_{g \in \mathfrak{g}} \delta(f,g) \leqslant n + 1}$$

Hence a meromorphic map $f : \mathbb{C}^m \longrightarrow \mathbb{P}(V)$ which misses $n + 2$ hyperplanes in general position is linearly degenerated (Borel [2]).

If $m = n = 1$, the defect relation was proved by Nevanlinna [9]. The case $m = 1 < n$ was solved by Cartan [3] and later Weyl [19], [20] and Ahlfors [1]. The First Main Theorem for $m > 1 = n$ was proved by H. Kneser [7]. The First Main Theorem and the Defect Relation for $m \geqslant 1 \leqslant n$ were established in [16].

Already Nevanlinna [10] wondered if his defect relation remained correct if $\mathfrak{g}$ is a finite set of meromorphic "target" maps $g : \mathbb{C}^m \longrightarrow \mathbb{P}(V^*)$ in general position with

(22) $$T_g(r,s)/T_f(r,s) \longrightarrow 0 \qquad \text{for } r \longrightarrow \infty$$

He verified the case $m = 1 = n$ and $\#\mathfrak{g} = 3$, which can be easily reduced to fixed targets. If $\mathfrak{g}$ is a finite set of polynomials of at most degree d on $\mathbb{C}$, Dufresnoy [6] proved a defect relation

(23) $$\sum_{g \in \mathfrak{g}} \delta(f,g) \leqslant 2 + d .$$

Also if $m = 1 = n$, if $\mathfrak{g}$ is a finite set of meromorphic functions satisfying (22), if $\mathfrak{g}$ spans a vector space of dimension p over $\mathbb{C}$, Chuang [5] proved

(24) $$\sum_{g \in \mathfrak{g}} \delta(f,g) \leqslant 1 + p(1 - \delta(f,\infty))$$

Hence he proved Nevanlinna's conjecture if f is entire on $\mathbb{C}$. If $m = 1 = n$, if f has finite lower order, Yang [22] showed that there are at most countably many meromorphic target functions g satisfying (22) with $\delta(f,g) > 0$. Recently, Osgood [11], [12] claimed that he can prove the Nevanlinna conjecture if $m = n = 1$ using number theory. However, at this writing, I know of no specialist in value distribution theory who has verified his proof.

If $m > 1 = n$ and if $\operatorname{Rank}(\mathfrak{g} \cup \{f\}) = 1 + \operatorname{Rank} \mathfrak{g}$, Shiffman [13], [14] proved the Nevanlinna conjecture, but for $m = 1$ his requirement forces the elements of $\mathfrak{g}$ to be constant. If $m = 1 = n$, S. Mori proves the Nevanlinna conjecture for the case $\#\mathfrak{g} = n + 2$ by reducing it to the case of fixed targets, see [10] and [17].

For $m \geqslant 1 \leqslant n$, a new value distribution theory for moving targets was developed in [17], which does not solve the Nevanlinna conjecture, but which contains the theorems of Shiffman and Mori. This theory and some of its improvements shall be outlined here in part.

A First Main Theorem shall be established for an abstract operator, which will be useful in a number of concrete situations. Let $V_1, \ldots, V_k$ and W be hermitian vector spaces. Assume that a k-linear map

$$\Theta : V_1 \times \ldots \times V_k \longrightarrow W \tag{25}$$

is given. We assume that Θ is not the zero map and preceive the map Θ as an "operation"

$$\mathfrak{x}_1 \Theta \ldots \Theta \mathfrak{x}_p = \Theta(\mathfrak{x}_1, \ldots, \mathfrak{x}_p) \tag{26}$$

for $\mathfrak{x}_j \in V$ if $j = 1, \ldots, k$. Take $x_j \in \mathbb{P}(V_j)$; then $x_j = \mathbb{P}(\mathfrak{x}_j)$ with $\mathfrak{x}_j \in V_{j*}$. Then $(\mathfrak{x}_1, \ldots, \mathfrak{x}_k)$ is said to be <u>free</u> for the operation Θ if and only if $\mathfrak{x}_1 \Theta \ldots \Theta \mathfrak{x}_k \neq 0$. Then

$$x_1 \Theta \ldots \Theta x_k = \mathbb{P}(\mathfrak{x}_1 \Theta \ldots \Theta \mathfrak{x}_k) \in \mathbb{P}(W) \tag{27}$$

is well defined. Also the Θ-distance

$$\square\, x_1 \dot{\Theta} \ldots \dot{\Theta} x_k \,\square = \frac{\|\mathfrak{x}_1 \Theta \ldots \Theta \mathfrak{x}_k\|}{\|\mathfrak{x}_1\| \ldots \|\mathfrak{x}_k\|} \tag{28}$$

is well-defined with $\square\, x_1 \dot{\Theta} \ldots \dot{\Theta} x_k \,\square > 0$ if and only if $(x_1, \ldots, x_k)$ is free for Θ. Here the dot over the operation symbol indicates that the distance is not a function of $x_1 \Theta \ldots \Theta x_k$ but of $(x_1, \ldots, x_k)$. By continuity

$$M_k(\Theta) = \operatorname{Max}\{\square\, x_1 \dot{\Theta} \ldots \dot{\Theta} x_k \,\square \mid x_j \in \mathbb{P}(V_j)\} > 0 \tag{29}$$

exists. The k-linear map Θ uniquely extends to a linear map

(30) $$\ominus : V_1 \otimes \dots \otimes V_k \longrightarrow W$$

with

(31) $$\ominus(\mathfrak{z}_1 \otimes \dots \otimes \mathfrak{z}_p) = \mathfrak{z}_1 \ominus \dots \ominus \mathfrak{z}_k .$$

Let $g_j : \mathbb{C}^m \longrightarrow \mathbb{P}(V_j)$ be meromorphic maps for $j = 1, \dots ,k$. Pick reduced representations $\mathfrak{w}_j : \mathbb{C}^m \longrightarrow V_j$ for g_j for $j = 1, \dots ,k$. Then $(g_1, \dots ,g_k)$ is said to be <u>free</u> for $\ominus$ if and only if

(32) $$\mathfrak{w} = \mathfrak{w}_1 \ominus \dots \ominus \mathfrak{w}_k \not\equiv 0 .$$

The condition does not depend on the choice of the representations. If $\mathfrak{w} \not\equiv 0$, then $\mathfrak{w}$ is a representation of a meromorphic map

(33) $$g = g_1 \ominus \dots \ominus g_k = \mathbb{P} \circ \mathfrak{w} : \mathbb{C}^m \longrightarrow W .$$

Also the <u>divisor</u>, <u>counting function</u> and <u>valence function</u> are well defined:

(34) $$\mu_{g\cdot} = \mu(g_1 \dot{\ominus} \dots \dot{\ominus} g_k) := \mu_{\mathfrak{w}} \geq 0$$

(35) $$n_{g\cdot} = n(g_1 \dot{\ominus} \dots \dot{\ominus} g_k) := n_{\mu_{\mathfrak{w}}} \geq 0$$

(36) $$N_{g\cdot} = N(g_1 \dot{\ominus} \dots \dot{\ominus} g_k) := N_{\mu_{\mathfrak{w}}} \geq 0 .$$

Since $\Box g_1 \dot{\ominus} \dots \dot{\ominus} g_k \Box \not\equiv 0$, the <u>compensation</u> <u>function</u>

(37) $$m_{g\cdot}(r) = m(r,g_1 \dot{\ominus} \dots \dot{\ominus} g_k) = \int_{\mathbb{C}^m\langle r\rangle} \log \frac{1}{\Box g_1 \dot{\ominus} \dots \dot{\ominus} g_k \Box} \sigma$$

is defined for all $r > 0$ with

(38) $$m_{g\cdot}(r) \geq \log \frac{1}{M_k(\ominus)} .$$

For $o < s < r$, we obtain the **First Main Theorem** (FMTH) for Θ

(39)
$$\sum_{j=1}^{k} T_{g_j}(r,s) = N_{g\cdot}(r,s) + m_{g\cdot}(r) - m_{g\cdot}(s) + T_g(r,s)$$

The dot $g\cdot$ indicates that the expression is not a functional of g. The characteristic T_g is an additional term in the FMTH, which in a sense corresponds to the deficit term of earlier FMTHs. However $N_{g\cdot}$ measures a divisor not a lower dimensional analytic set as in the case in FMTHs with deficits. If $\dim W = 1$, then $\dim \mathbb{P}(W) = 0$ and the map g is constant which implies $T_g(r,s) \equiv 0$.

If at least one map g_j is not constant $\sum_{j=1}^{k} T_{g_j}(r,s) \longrightarrow \infty$ for $r \longrightarrow \infty$ and the **defect** is defined by

(40)
$$\delta(g\cdot) = \delta(g_1 \dot{\Theta} \dots \dot{\Theta} g_k) = \underline{\lim}_{r\to\infty} \frac{m_{g\cdot}(r)}{T_{g_1}(r,s) + \dots + T_{g_k}(r,s)}$$

The First Main Theorem implies

(41)
$$0 \leqslant \delta(g\cdot) = 1 - \overline{\lim}_{r\to\infty} \frac{N_{g\cdot}(r,s) + T_{g\cdot}(r,s)}{T_{g_1}(r,s) + \dots + T_{g_p}(r,s)} \leqslant 1$$

If $\mathbb{C}^m$ were replaced by a compact Kähler manifold, (39) would be replaced by an identity between cohomology classes without boundary terms. Thus the defect would be zero. In the open case, the boundary term compensates for the difference in the form identities, where the forms represent the classes. The defect measures the difference. In (41), the defect becomes positive, if the operation Θ binds the maps $g_1, \dots, g_k$ only weakly compared to the unbound growth of the $g_1, \dots, g_k$. A defect relation would restrict the number of such cases among a reasonable system of maps $(g_1, \dots, g_k)$.

The First Main Theorem can be applied to a number of operations the exterior product $\wedge$, the interior product $\llcorner$, the inner product $\langle \, , \, \rangle$ and the connection product $\boxplus$ and its iteration $\boxplus_\rho$.

1. The exterior product and general position. Let V a hermitian vector space of dimension $n+1>1$. Take $V_j = V^*$ for $j = 1, \dots, k$ with $1 \leqslant k \leqslant n+1$. Take $\Theta = \wedge$ and $W = \bigwedge_k V^*$. Let $g_j : \mathbb{C}^m \longrightarrow \mathbb{P}(V^*)$ be meromorphic maps and assume that $(g_1, \dots, g_k)$ is free for $\wedge$. Then

(42)
$$T_{g_1}(r,s) + \dots + T_{g_k}(r,s) - T_{g_1 \wedge \dots \wedge g_k}(r,s)$$
$$= N(r,s,g_1 \ddot{\wedge} \dots \ddot{\wedge} g_k) + m(r,g_1 \ddot{\wedge} \dots \ddot{\wedge} g_k) - m(s,g_1 \ddot{\wedge} \dots \ddot{\wedge} g_k) .$$

If $k = n+1$, then $T_{g_1 \wedge \dots \wedge g_{n+1}} \equiv 0$. This First Main Theorem allows the treatment of general position.

A family $\mathfrak{g} = (g_j)_{j=1,\dots,k}$ is said to be in general position, if there is a point $\mathfrak{z}_0 \in \mathbb{C}^m$ such that all the meromorphic maps $g_j : \mathbb{C}^m \longrightarrow \mathbb{C}(V^*)$ are holomorphic at $\mathfrak{z}_0$, such that $g_1(\mathfrak{z}_0), \dots, g_k(\mathfrak{z}_0)$ are pair wise different and such that $\{g_1(\mathfrak{z}_0), \dots, g_k(\mathfrak{z}_0)\}$ is in general position. If $k < n+1$ and if $\mathfrak{g}$ is in general position, then $(g_1, \dots, g_k)$ is free for $\wedge$ (and only then). We define

(43)
$$\Gamma_{\mathfrak{g}}(r) = m(r,g_1 \ddot{\wedge} \dots \ddot{\wedge} g_k)$$

and the First Main Theorem yields the estimate

(44)
$$\Gamma_{\mathfrak{g}}(r) \leqslant \sum_{j=1}^{k} T_{g_j}(r,s) + m(s,g_1 \ddot{\wedge} \dots \ddot{\wedge} g_k) .$$

Now, consider the case $k \geqslant n+1$. Let $\mathfrak{J}(n+1,k)$ be the set of all injective increasing maps $j : \mathbb{N}[1,n+1] \longrightarrow \mathbb{N}[1,k]$. At every point $\mathfrak{z} \in \mathbb{C}^m$ where all the maps g_j are holomorphic define

(45)
$$\Gamma_{\mathfrak{g}}(\mathfrak{z}) = \mathrm{Min}\{\square\, g_{j(1)}(\mathfrak{z}) \ddot{\wedge} \dots \ddot{\wedge} g_{j(n+1)}(\mathfrak{z})\, \square \mid j \in \mathfrak{J}(n+1,k)\} .$$

Then $\Gamma_{\mathfrak{g}}(\mathfrak{z}) > 0$ for all $\mathfrak{z}$ outside a thin analytic subset of $\mathbb{C}^m$ if and only if $\mathfrak{g}$ is in general position. The gauge function $\Gamma_{\mathfrak{g}}$ is now defined by

(46) $$\Gamma_{\mathfrak{g}}(r) = \int_{\mathbb{C}^m\langle r\rangle} \log \frac{1}{\Gamma_{\mathfrak{g}}} \sigma \geqslant 0 \qquad \text{for all} \quad r > 0 .$$

The First Main Theorem is applied to $(g_{j(1)}, \ldots, g_{j(n+1)})$ for each $j \in \mathfrak{J}(n+1,k)$. Addition yields

(47) $$0 \leqslant \Gamma_{\mathfrak{g}}(r) \leqslant \binom{k-1}{n} \sum_{j=1}^{k} T_{g_j}(r,s) + c(s)$$

where $c(s)$ is a constant.

2. **The connection operator**. Let V be a hermitian vector space of dimension $n + 1 > 1$. Take integers p and q with $0 \leqslant p \leqslant n$ and $0 \leqslant q \leqslant n$. Define $\mu = \mathrm{Min}(p+1,q+1)$. Then there exists one and only one bilinear operation

(48) $$\boxplus : (\textstyle\bigwedge_{q+1} V) \times (\bigwedge_{p+1} V^*) \longrightarrow (\bigwedge_q V) \otimes (\bigwedge_p V^*)$$

which satisfies the following property: Take $\mathfrak{x}_j \in V$ for $j = 0,1, \ldots ,q$ and $\mathfrak{y}_k \in V$ for $k = 0,1, \ldots ,p$. Put

(49) $$\mathfrak{x} = \mathfrak{x}_0 \wedge \cdots \wedge \mathfrak{x}_q \qquad \mathfrak{y} = \mathfrak{y}_0 \wedge \cdots \wedge \mathfrak{y}_p$$

(50) $$\hat{\mathfrak{x}}_j = (-1)^j \mathfrak{x}_0 \wedge \cdots \wedge \mathfrak{x}_{j-1} \wedge \mathfrak{x}_{j+1} \wedge \cdots \wedge \mathfrak{x}_q$$

(51) $$\hat{\mathfrak{y}}_k = (-1)^k \mathfrak{y}_0 \wedge \cdots \wedge \mathfrak{y}_{k-1} \wedge \mathfrak{y}_{k+1} \wedge \cdots \wedge \mathfrak{y}_p$$

Then

(52) $$\mathfrak{x} \boxplus \mathfrak{y} = \frac{1}{\mu} \sum_{j=0}^{q} \sum_{k=0}^{p} \langle \mathfrak{x}_j, \mathfrak{y}_k \rangle \hat{\mathfrak{x}}_j \otimes \hat{\mathfrak{y}}_k .$$

The **Grassmann cone** of order q is given by

(53) $$\tilde{G}_q(V) = \{\mathfrak{x}_0 \wedge \cdots \wedge \mathfrak{x}_q \mid \mathfrak{x}_j \in V\}$$

and $G_q(V) = \mathbb{P}(\tilde{G}_q(V))$ is the **Grassmann manifold** of order q. If $x \in G_q(V)$, then $x = \mathbb{P}(\mathfrak{x})$ with $\mathfrak{x} = \mathfrak{x}_0 \wedge \cdots \wedge \mathfrak{x}_q \neq 0$ and

$$(54) \qquad E(x) = \mathbb{P}(\mathbb{C}\mathfrak{x}_0 + \ldots + \mathbb{C}\mathfrak{x}_q)$$

is a q-dimensional projective plane in $\mathbb{P}(V)$. If $y \in G_q(V^*)$, then $y = \mathbb{P}(\mathfrak{y})$ with $\mathfrak{y} = \mathfrak{y}_0 \wedge \ldots \wedge \mathfrak{y}_p \neq 0$ and

$$(55) \qquad E[y] = \mathbb{P}(\ker \mathfrak{y}_0 \cap \ldots \cap \ker \mathfrak{y}_p)$$

is a $(n-p-1)$-dimensional projective plane in $\mathbb{P}(V)$. If $p + q + 1 \leqslant n$, then $\mathfrak{x} \boxplus \mathfrak{y} = 0$ if and only if $E(x) \subseteq E[y]$. If $x \in G_q(V)$ and $y \in G_p(V^*)$ then

$$(56) \qquad \square\, x \,\dot{\boxplus}\, y \,\square \leqslant \frac{1}{\sqrt{\mu}} .$$

The Ta-operator (48) extends to a linear map

$$(57) \qquad \boxplus : (\bigwedge_{q+1} V) \otimes (\bigwedge_{p+1} V^*) \longrightarrow (\bigwedge_q V) \otimes (\bigwedge_p V^*)$$

which can be iterated $\boxplus_\rho = \boxplus_{\rho-1} \circ \boxplus$ if $0 < \rho \leqslant \mu$. If $\rho = \mu$, then $\boxplus_\rho$ becomes the interior product $\llcorner$ and if $\rho = p + 1 = q + 1$, then $\boxplus_\rho$ becomes the inner product $\langle\, ,\, \rangle$. If $x \in G_q(V)$ and $y \in G_p(V^*)$ with $p + q + 1 \leqslant n$ and if $1 \leqslant \rho \leqslant \mu$, then

$$(58) \qquad \square\, x \,\dot{\boxplus}_\rho\, y \,\square \leqslant \frac{1}{\sqrt{\binom{\mu}{\rho}}} .$$

Moreover $\square\, x \,\dot{\boxplus}_\rho\, y \,\square = 0$ if and only if $\dim E(x) \cap E[y] > q - \rho$. If $f : \mathbb{C}^m \longrightarrow G_q(V)$ and $g : \mathbb{C}^m \longrightarrow G_p(V^*)$ are meromorphic maps such that (f,g) is free for $\boxplus_\rho$, the First Main Theorem reads

$$(59) \qquad T_f(r,s) + T_g(r,s) = N(r,s,f \,\dot{\boxplus}_\rho\, g) + m(r,f \,\dot{\boxplus}_\rho\, g) - m(s,f \,\dot{\boxplus}_\rho\, g) + T_{f\boxplus_\rho g}(r,s) .$$

If $\rho = 1$, we write $\boxplus_1 = \boxplus$. If $q + 1 \geqslant p + 1 = \rho$, then $f \,\dot{\boxplus}_\rho\, g = f \,\dot{\llcorner}\, g$.

If $p + 1 \geqslant q + 1 = \rho$ then $f \,\dot{\boxplus}_\rho\, g = g \,\dot{\llcorner}\, f$. If $p + 1 = q + 1 = \rho$, then

$f \dot{\boxplus}_\rho g = <f;g> = f,g$. The First Main Theorem shall be applied to the associated maps which shall be defined now.

Take a holomorphic exterior differential form

$$B = \sum_{j=1}^{m} B_j(-1)^{j-1}\, dz_1 \wedge \ldots \wedge dz_{j-1} \wedge dz_{j+1} \wedge \ldots \wedge dz_m \tag{60}$$

of bidegree $(m-1,0)$ on $\mathbb{C}^m$. For any holomorphic vector function $\mathfrak{v} : \mathbb{C}^m \longrightarrow V$ define

$$\mathfrak{v}' = \sum_{j=1}^{m} B_j \frac{\partial \mathfrak{v}}{\partial z_j} \quad \text{that is} \quad d\mathfrak{v} \wedge B = \mathfrak{v}' dz_1 \wedge \ldots \wedge dz_m . \tag{61}$$

The operation can be iterated $\mathfrak{v}^{(q)} = (\mathfrak{v}^{(q-1)})'$. Let $f : \mathbb{C}^m \longrightarrow \mathbb{P}(V)$ be a meromorphic map. Let $\mathfrak{v} : \mathbb{C}^m \longrightarrow V$ be a reduced representation. For $0 \leqslant q \in \mathbb{Z}$ a holomorphic map

$$\mathfrak{v}_q = \mathfrak{v} \wedge \mathfrak{v}' \wedge \ldots \wedge \mathfrak{v}^{(q)} : \mathbb{C}^m \longrightarrow G_q(V) \tag{62}$$

is defined. One and only one number $\ell_f = \ell_f(B)$, called the **index** of **generality** of f for B exists such that $\mathfrak{v}_q \not\equiv 0$ if $0 \leqslant q \leqslant \ell_f$ and such that $\mathfrak{v}_q \equiv 0$ if $q > \ell_f$. Naturally, $0 \leqslant \ell_f \leqslant n$. If f is linearly non-degenerated, then there exists a holomorphic form B of bidegree $(m-1,0)$ on $\mathbb{C}^m$ with polynomial coefficients of at most degree $n-1$, such that $\ell_f(B) = n$. If $0 \leqslant q \leqslant \ell_f$, then $\mathfrak{v}_q$ is a representation of a meromorphic map $f_q : \mathbb{C}^m \longrightarrow G_q(V)$ called the qth **associated map** of f. Moreover the non-negative divisor $F_q = \mu_{\mathfrak{v}_q}$ does not depend on the choice of $\mathfrak{v}$. If $0 \leqslant q < \ell_f$, the so called **stationary divisor**

$$\iota_q = F_{q-1} - 2F_q + F_{q+1} \geqslant 0 \tag{63}$$

is non-negative. For every integer $a \geqslant 0$ define

$$i_a = a!(-1)^{\frac{a(a-1)}{2}} \left[\frac{i}{2\pi}\right]^a . \tag{64}$$

Let Ω_q be the Fubini-Study form on $G_q(V)$. Define

$$\mathbb{H}_q = m i_{m-1}\, f_q^*(\Omega_q) \wedge B \wedge \bar{B} \geqslant 0 \tag{65}$$

If $0 \leqslant q < \ell_f$, then $\mathbb{H}_q > 0$ outside a thin analytic subset of $\mathbb{C}^m$. Also $\mathbb{H}_{\ell_f} \equiv 0$.

For the remainder of these notes, we shall assume that $\ell_f = n$, which implies that ℓ_f is linearly non-degenerate.

Let $g : \mathbb{C}^m \longrightarrow \mathbb{P}(V^*)$ be a meromorphic map. For $0 \leqslant p \leqslant \ell_g$ the associated maps $g_p : \mathbb{C}^m \longrightarrow G_p(V^*)$ are defined. Let Ω_p^* be the associated Fubini Study form on $G_p(V^*)$. Denote

(66)
$$\mathbb{K}_p = \mathbb{K}_p(g) = mi_{m-1} g_p^*(\Omega_p^*) \wedge B \wedge \bar{B} \geqslant 0$$

where $\mathbb{K}_{\ell_g} \equiv 0$ and $\mathbb{K}_p > 0$ outside a thin analytic subset of $\mathbb{C}^m$ if $0 \leqslant p < \ell_g$. For $0 \leqslant q \leqslant n$ and $0 \leqslant p \leqslant \ell_g$ and $0 \leqslant \rho \leqslant \mu = \mathrm{Min}(p+1, q+1)$ define

(67)
$$\Phi_{pq}^{(\rho)} := \begin{bmatrix} \mu \\ \rho \end{bmatrix}^2 \square f_q \,\dot{\boxplus}_\rho\, g_p \square^2$$

Then $\Phi_{pq}^{(\rho)} \leqslant \begin{bmatrix} \mu \\ \rho \end{bmatrix}$. Here $\Phi_{pq}^{(\rho)} \not\equiv 0$ if and only if (f_q, g_p) is free for $\boxplus_\rho$. If $\rho = 1$, abbreviate $\Phi_{pq} := \Phi_{pq}^{(1)}$.

Important is the __Stress__ __Curvature__ __Formula__:

(68)
$$\begin{aligned} & mi_{m-1}\, dd^c \Phi_{pq}^{(\rho)} \wedge B \wedge \bar{B} \\ & = (\Phi_{p+1,q}^{(\rho)} - 2\Phi_{pq}^{(\rho)} + \Phi_{p-1,q}^{(\rho)})\mathbb{K}_p \\ & + (\Phi_{p,q-1}^{(\rho)} - 2\Phi_{pq}^{(\rho)} + \Phi_{p-1,q}^{(\rho)})\mathbb{H}_q \\ & - 2S(p,q,\rho)\sqrt{\mathbb{K}_p \mathbb{H}_q} \end{aligned}$$

The twisting term $S(p,q,\rho)$ will be defined shortly. The identity (67) looks simple and symmetric but its proof requires a gigantic calculation in particular if $\rho > 1$. The case $\rho = 1$ is published in [17]. The Stress Curvature Formula is essential in the proof of the Second Main Theorem. Since (67) governs the relation between the jets of the two maps f and g, the formula may be of interest outside value distribution theory.

The term $S(p,q,\rho)$ will be explained first if $\rho = 1$. Take reduced representations $\mathfrak{w} : \mathbb{C}^m \longrightarrow V$ of f and $\mathfrak{v} : \mathbb{C}^m \longrightarrow V^*$ of g. At every point outside a thin analytic set $\mathfrak{w}, \mathfrak{w}', \dots, \mathfrak{w}^{(n)}$ is a base of V which we orthonormalize to $\mathfrak{e}_0, \mathfrak{e}_1, \dots, \mathfrak{e}_n$. Let $\mathfrak{e}_0^*, \dots, \mathfrak{e}_n^*$ be the dual frame in V^*. At every point outside a thin analytic set $\mathfrak{v}, \mathfrak{v}', \dots, \mathfrak{v}^{(\ell_g)}$ are linearly independent in V^* and are orthonormalized to $\psi_0, \psi_1, \dots, \psi_{\ell_g}$. For $p \in \mathbb{Z}[0,\ell_g]$ and $q \in \mathbb{Z}[0,n]$ there are unique functions Λ_{pq} on the complement of a thin analytic subset of $\mathbb{C}^m$ such that

(69)
$$\psi_p = \sum_{q=1}^{n} \Lambda_{pq} \mathfrak{e}_q^* \qquad \text{for all } p \in \mathbb{Z}[0,\ell_g] .$$

For $p \in \mathbb{Z}[0,\ell_g]$, $q \in \mathbb{Z}[0,n]$, $a \in \mathbb{Z}[0,\ell_g]$ and $b \in \mathbb{Z}[0,n]$ define

(70)
$$S_{pq}^{ab} = \frac{1}{2}(\Lambda_{pq}\bar{\Lambda}_{ab} + \Lambda_{ab}\bar{\Lambda}_{pq}) = S_{ab}^{pq} .$$

Then $-1 \leqslant S_{pq}^{ab} \leqslant 1$. If $a + b = p + q$, then S_{pq}^{ab} is a global invariant independent of the choice of the representations $\mathfrak{w}$, $\mathfrak{v}$ and the coordinate system $z_1, \dots, z_m$ on $\mathbb{C}^m$ even locally. Then

(71)
$$S(p,q,1) = S_{p+1,\, q}^{p\,,\,q+1}$$

(72)
$$\Phi_{pq} = \sum_{j=0}^{p} \sum_{k=0}^{q} S_{jk}^{jk} .$$

For $\rho > 1$, the basic idea is the same, but the situation is more complicated. If a and b are integers with $0 \leqslant a \leqslant b$, let $\mathfrak{F}[a,b]$ be the set of all increasing, injective maps $\alpha : \mathbb{Z}[0,a] \longrightarrow \mathbb{Z}[0,b]$. If $c \in \mathbb{Z}$ with $c > b$, then $\beta = (\alpha,c)$ is the unique map $\beta \in \mathfrak{F}[a + 1,c]$ where $\beta \mid \mathbb{Z}[0,a] = \alpha \in \mathfrak{F}[a,b]$ and $\beta(a + 1) = c$. If $\alpha \in \mathfrak{F}[\rho - 1,\ell_g]$ and $\beta \in \mathfrak{F}[\rho - 1,n]$, define the stress coefficient

$$\text{(73)} \qquad \Lambda_{\alpha\beta} = \begin{vmatrix} \Lambda_{\alpha(0)\beta(0)}, & \cdots & ,\Lambda_{\alpha(0)\beta(\rho-1)} \\ \cdot & & \cdot \\ \cdot & & \cdot \\ \cdot & & \cdot \\ \cdot & & \cdot \\ \cdot & & \cdot \\ \Lambda_{\alpha(\rho-1),\beta(0)}, & \cdots & ,\Lambda_{\alpha(\rho-1),\beta(\rho-1)} \end{vmatrix}$$

Then $|\Lambda_{\alpha\beta}| \leqslant 1$ and

$$\text{(74)} \qquad \Phi_{pq}^{(\rho)} = \sum_{\alpha\in \mathfrak{P}[\rho-1,p]} \sum_{\beta\in \mathfrak{P}[\rho-1,q]} |\Lambda_{\alpha\beta}|^2 .$$

If $\alpha \in \mathfrak{P}[a,b]$, define $\vec{\alpha} = \alpha(0) + \ldots + \alpha(a)$. If $\alpha \in \mathfrak{P}[\rho - 1,\ell_g]$, $\beta \in \mathfrak{P}[\rho - 1,n]$, $\gamma \in \mathfrak{P}[\rho - 1,\ell_g]$ and $\delta \in \mathfrak{P}[\rho - 1,n]$ with $\vec{\alpha} + \vec{\beta} = \vec{\gamma} + \vec{\delta}$, then

$$\text{(75)} \qquad S_{\alpha\beta}^{\gamma\delta} = \tfrac{1}{2}(\Lambda_{\alpha\beta}\bar{\Lambda}_{\gamma\delta} + \Lambda_{\gamma\delta}\bar{\Lambda}_{\alpha\beta}) = S_{\gamma\delta}^{\alpha\beta}$$

is a global invariant independent of the choice of the representation $\mathfrak{v}$ and $\mathfrak{w}$ and the coordinate systems on $\mathbb{C}^m$ even locally. If $p \in \mathbb{Z}[0,\ell_g]$, if $q \in \mathbb{Z}[0,n]$ if $\mu = \mathrm{Min}(p+1,q+1)$ and if $\rho \in \mathbb{Z}[0,\mu]$, then

$$\text{(76)} \qquad S(p,q,\rho) = \sum_{\alpha\in \mathfrak{P}[\rho-1,p-1]} \sum_{\alpha\in \mathfrak{P}[\rho-,q-1]} S^{(\alpha,\ p\ ),(\beta,q+1)}_{(\alpha,p+1),(\beta,\ q\ )} .$$

For $p \in \mathbb{Z}[0,\ell_g]$ and $q \in \mathbb{Z}[0,n]$ define

$$0 \leqslant \Xi_{pq}^{\rho} = \Phi_{p+1,q}^{(\rho)}\mathbb{K}_p + \Phi_{p,q+1}^{(\rho)}\mathbb{H}_q + 2S(p,q,\rho)\sqrt{\mathbb{K}_p\mathbb{H}_q} .$$

For $r > 0$ define

$$\text{(77)} \qquad \tilde{Y}(r) = \mathrm{Max}\Big(\sum_{j=1}^{m} |B_j(\mathfrak{z})|^2 \,\Big|\, \mathfrak{z} \in \mathbb{C}^m[r]\Big)$$

and $Y(r) = \mathrm{Max}(1,\tilde{Y}(r))$. Assume that (f_q,g_p) is free for $\mathbb{H}_\rho$. Take real numbers β, s, r with $0 \leqslant \beta \leqslant 1$ and $0 < s < r$. Then the **Ahlfors Estimates** hold

(78)
$$\beta^2 \int_s^r \int_{\mathbb{C}^m[t]} \square f_q \dot{\boxplus}_\rho g_p \square^{2\beta-2} \Xi_{pq}^{(\rho)} \frac{dt}{t^{2m-1}}$$

$$\leqslant 4(n+1)^{2\rho+1} Y(r)(T_{f_q}(r,s) + T_{g_p}(r,s) + 1) .$$

For $r > 0$ define the <u>**obstruction terms**</u>

(79)
$$P^{\rho}_{pq}(r,g) = \frac{1}{2} \int_{\mathbb{C}^m\langle r\rangle} \log \frac{\Phi^{(\rho)}_{p,q+1} \mathbb{H}_q}{\Xi^{\rho}_{pq}} \sigma$$

(80)
$$R_{pq}(r,g) = \int_{\mathbb{C}^m\langle r\rangle} \log(1 + \sqrt{\mathbb{K}_p/\mathbb{H}_q})\sigma \geqslant 0 .$$

If $p = \ell_g$, then $\mathbb{K}_p = 0$ and $P^{\rho}_{pq} \equiv 0 \equiv R_{pq}$. Abbreviate $P^1_{pq} = P_{pq}$.

The Stress Curvature Formula, the Ahlfors estimates and the theory of general position lead to the <u>**Second Main Theorem**</u>.

Let $\mathfrak{G}$ be a finite set of meromorphic maps $g : \mathbb{C}^m \longrightarrow \mathbb{P}(V^*)$. Put $k = \#\mathfrak{G}$. Take $p \in \mathbb{Z}$ with $0 \leqslant p \leqslant \ell_g$ for all $g \in \mathfrak{G}$. Define $\mathfrak{G}_p = \{g_p \mid g \in \mathfrak{G}\}$. Since $g_p : \mathbb{C}^m \longrightarrow \mathbb{P}(\bigwedge_{p+1} V^*)$ it makes sense to assume that $\mathfrak{G}_p$ is in general position. If $k < \begin{pmatrix} n+1 \\ p+1 \end{pmatrix}$, then (44) implies

(81)
$$\Gamma_{\mathfrak{G}_p}(r) \leqslant \sum_{g \in \mathfrak{G}} T_{g_p}(r,s) + c_o(s)$$

If $k \geqslant \begin{pmatrix} n+1 \\ p+1 \end{pmatrix}$, then (47) implies

(82)
$$\Gamma_{\mathfrak{G}_p}(r) \leqslant \left(\frac{k-1}{\begin{pmatrix} n+1 \\ p+1 \end{pmatrix} - 1} \right) \sum_{g \in \mathfrak{G}} T_{g_p}(r,s) + c_1(s)$$

For simplicity, we shall formulate the Second Main Theorem for the case $\rho = 1$ first. Take $q \in \mathbb{Z}[0,n-p-1]$. Assume that (f_q,g_p) is free for $\boxplus$. Take $\epsilon > 0$ and $s > 0$.

Then there exists a set E of finite measure on the positive real axis $\mathbb{R}^+$ such that the **Second Main Theorem** holds for all $r \in \mathbb{R}^+ - E$.

$$\sum_{h=q}^{n-p-1} \begin{bmatrix} n-h \\ p+1 \end{bmatrix} N_{\ell_h}(r,s) + \sum_{h=q+1}^{n-p-1} \begin{bmatrix} n-h-1 \\ p-1 \end{bmatrix} T_{f_h}(r,s)$$

$$+ \sum_{g \in \mathfrak{G}} m(r, f_q \dot{\boxplus} g_p)$$

$$\text{(83)} \qquad \leq \left(\begin{bmatrix} n-q \\ p+1 \end{bmatrix} + \begin{bmatrix} n-q-1 \\ p \end{bmatrix} \right) T_{f_g}(r,s) + \sum_{h=q}^{n-p-1} \sum_{g \in \mathfrak{G}} P_{ph}(r,g)$$

$$+ \sum_{h=q}^{n-p-1} \left(k - \begin{bmatrix} n-h \\ p+1 \end{bmatrix} \right)^+ (\Gamma_{\mathfrak{G}_p}(r) + \sum_{g \in \mathfrak{G}} R_{ph}(r,g)) + \epsilon \log r$$

$$+ 3(1+\epsilon) \begin{bmatrix} n+1 \\ p+2 \end{bmatrix} (k \log T_{f_q}(r,s) + 2k \log Y(r) + \sum_{g \in \mathfrak{G}} \log T_{g_p}(r,s))$$

Therefore if for $r \longrightarrow \infty$ we assume

$$\text{(84)} \qquad T_{g_p}(r,s)/T_{f_q}(r,s) \longrightarrow 0 \qquad \text{for all } g \in \mathfrak{G}$$

$$\text{(85)} \qquad \log Y(r)/T_{f_q}(r,s) \longrightarrow 0$$

$$\text{(86)} \qquad R_{ph}(r,g)/T_{f_q}(r,s) \longrightarrow 0 \qquad \text{for } h \in \mathbb{Z}[q, n-p-1] \text{ and } g \in \mathfrak{G}$$

$$\text{(87)} \qquad P_{ph}(r,g)/T_{f_q}(r,s) \longrightarrow 0 \qquad \text{for } h \in \mathbb{Z}[q, n-p-1] \text{ and } g \in \mathfrak{G}$$

we obtain the **Defect Relation**

$$\text{(88)} \qquad \boxed{\sum_{g \in \mathfrak{G}} \delta(f_q \dot{\boxplus} g_p) \leq \begin{bmatrix} n-q \\ p+1 \end{bmatrix} + \begin{bmatrix} n-q-1 \\ p \end{bmatrix}} \,.$$

(84) is the natural extension of Nevanlinna's assumption. If B is chosen with polynomial coefficients and if f_q is transcendental, then (85) is satisfied. If $p = \ell_g$ for all $g \in \mathfrak{G}$, then $R_{ph} \equiv 0 \equiv P_{ph}$ and (86) and (87) are satisfied. In particular cases we obtain

$$\text{(89)} \qquad \sum_{g \in \mathfrak{G}} \delta(f_q \dot{\llcorner} g) \leq n - q - 1 \qquad \text{if } p = 0$$

(90) $$\sum_{g \in \mathfrak{G}} \delta(g_p \dot{\perp} f) \leqslant \binom{n}{p+1} + \binom{n-1}{p} \qquad \text{if } q = 0$$

(91) $$\sum_{g \in \mathfrak{G}} \delta(f,g) \leqslant n + 1 \qquad \text{if } p = q = 0$$

(92) $$\sum_{g \in \mathfrak{G}} \delta(f_q \dot{\boxplus} g_{n-q-1}) \leqslant 2 \qquad \text{if } p + q + 1 = n$$

For (90) see [17]. (91) is Nevanlinna's conjecture valid under the stated conditions. The incidence $E(f_q(z)) \subseteq E[g_p(z)]$ studied in (88) becomes the incidence $E(f_q(z)) = E[g_p(z)]$ in (92) where $p + q + 1 = n$ with the classical defect bound 2.

Now we shall consider the case $\rho > 1$. For notational purposes we replace ρ by $\rho + 1$ such that we are considering the case $\rho > 0$. Again let $\mathfrak{G}$ be a finite set of meromorphic maps $g : \mathbb{C}^m \longrightarrow \mathbb{P}(V^*)$. Put $k = \#\mathfrak{G}$. Take $p \in \mathbb{Z}$ with $0 \leqslant p \leqslant \ell_g$ for all $g \in \mathfrak{G}$. Assume that $\mathfrak{G}_p = \{g_p \mid g \in \mathfrak{G}\}$ is in general position. Then (81) respectively (82) holds. For each $x \in \mathbb{R}$ let $x^+ = \mathrm{Max}(0,x)$. Let ρ and q be integers with

(93) $$0 < \rho \leqslant p < n \qquad \rho \leqslant q \leqslant n + \rho - p - 1 \qquad p + q + 1 \leqslant n + \rho$$

Define

(94) $$\mathfrak{A}(p,q,\rho) = \{h \in \mathbb{Z}[q + 1, n + \rho - p - 2] \mid p(h + 1) < \rho(n + 1)\}$$

(95) $$\mathfrak{B}(p,q,\rho) = \{h \in \mathbb{Z}[q + 1, n + \rho - p - 2] \mid p(h + 1) > \rho(n + 1)\}$$

(96) $$Q = \mathrm{Max}\, \mathfrak{A}(p,q,\rho) \qquad \text{if } \mathfrak{A}(p,q,\rho) \neq \emptyset$$

(97) $$Q = q \qquad \text{if } \mathfrak{A}(p,q,\rho) = \emptyset$$

(98) $$A(p,h,\rho) = \binom{h}{\rho}\binom{n-h}{p-\rho} - \binom{h+1}{\rho}\binom{n-h-1}{p-\rho}$$

Then $A(p,h,\rho) < 0$ if $h \in \mathfrak{A}(p,q,\rho)$ and $A(p,h,\rho) > 0$ if $h \in \mathfrak{B}(p,q,\rho)$ and $A(p,h,\rho) = 0$ if $(h + 1)p = \rho(n + 1)$. Define

(99) $$D(p,h,\rho) = \sum_{j=0}^{\rho} \binom{h+1}{j}\binom{n-h}{p+1-j}$$

(100) $$0 \leq B(p,q,\rho) = D(p,q,\rho) + \binom{Q+1}{\rho}\binom{n-q-1}{p-\rho}(Q-q) - \sum_{h=q+1}^{Q}\binom{h}{\rho}\binom{n-h}{p-\rho}$$

(101) $$0 \leq c(p,q,\rho) = D(p,q,\rho) + \binom{Q+1}{\rho}\binom{n-Q-1}{p-\rho} - \sum_{h=q+1}^{Q}\binom{h}{\rho}\binom{n-h}{p-\rho} + \binom{n+\rho-p}{\rho}\frac{n-p+\rho-q}{n-p+\rho}((n-p+1)\rho-(n-p)p)^{+}$$

(102) $$L(p,\rho) = \binom{n+\rho-p-1}{\rho}(p+1-\rho) - \binom{n+\rho-p}{\rho}$$

For $r > 0$ abbreviate

(103) $$\mathbb{R}_{\rho}(r) = \sum_{h=q}^{n+\rho-p-1}(k - D(p,h,\rho))^{+}(\Gamma_{\mathfrak{G}_p}(r) + \sum_{g\in\mathfrak{G}} R_{ph}(r,g)) + \sum_{h=q}^{n+\rho-p-1}\sum_{g\in\mathfrak{G}} P_{ph}^{(\rho+1)}(r,g) + \lambda(1+\epsilon)(k \log T_{f_q}(r,s) + 2k \log Y(r) + \sum_{g\in\mathfrak{G}} \log T_{g_p}(r,s))$$

where λ is a constant which depends on n, p, q, ρ, Q only. We assume that (f_q, g_p) is free for $\boxplus_{\rho+1}$. We take $\epsilon > 0$ and $s > 0$. Then there exists a set E of finite measure in $\mathbb{R}^+$ such that the <u>**Second Main Theorem**</u> holds for all $r \in \mathbb{R}^+ - E$.

(104)

$$\sum_{h=q}^{n+\rho-p-1} D(p,h,\rho)N_{\iota_h}(r,s) + \sum_{h\in\mathfrak{F}(p,q,\rho)} A(p,h,\rho)T_{f_h}(r,s) + B(p,q,\rho)T_{q-1}(r,s) + L(p,\rho)^{+}T_{f_{n+\rho-p-1}}(r,s) + \sum_{g\in\mathfrak{G}} m(r,f_q \dot{\boxplus}_{\rho+1} g_p) \leq c(p,q,\rho)T_{f_q}(r,s) + \mathbb{R}_{\rho}(r) + \epsilon \log r$$

If we assume that (84) and (85) hold and if we assume

(86_ρ) $\quad R_{ph}(r,g)/T_{f_q}(r,s) \longrightarrow 0$ $\quad$ for $h \in \mathbb{Z}[q,n-p-1+\rho]$ and for $g \in \mathfrak{g}$

(87_ρ) $\quad P_{ph}^{(\rho+1)}(r,g)/T_{f_g}(r,s) \longrightarrow 0$ $\quad$ for $h \in \mathbb{Z}[q,n-p-1+\rho]$ and for $g \in \mathfrak{g}$

if $r \longrightarrow \infty$, then we obtain the <u>**Defect**</u> <u>**Relation**</u>

(105)
$$\boxed{\sum_{g \in \mathfrak{g}} m(r, g_q \,\dot{\boxplus}_{\rho+1}\, g_p) \leq c(p,q,\rho)}$$

Recall that the defect in (105) is related to the incidence

(106)
$$\dim E(f_q(\mathfrak{z})) \cap E[g_p(\mathfrak{z})] > q - \rho - 1,$$

a condition which represents Schubert cycle in $G_q(V)$. We may ask if there are defect relations connected with any Schubert cycle conditions in $G_q(V)$.

The true meaning of the various terms in (104) and of the defect bound $c(p,q,\rho)$ in (105) remain to be explored. For this report, $\mathbb{C}^m$ served as the domain space for the maps. However the theory carries over to parabolic manifolds (M,τ). Naturally, there are not necessarily global representations available, but this is only a minor difficulty. In the case of a parabolic manifold, the Ricci function of the manifold $\mathrm{Ric}_\tau(r,s)$ appears, which for branch coverings $\pi : M \longrightarrow \mathbb{C}^m$ is the valence function of the branching divisor.

References

[1] L. Ahlfors, The theory of meromorphic curves, Acta Soc. Sci. Fenn, Nova Ser. A 3(4)(1941), 171-183.

[2] E. Borel, Sur les zéros des fonctions entières., Acta Math., 20(1897), 357-396.

[3] H. Cartan, Sur les zéros des combinaisons linéaires de p fonctions holomorphes données, Mathematica (Cluj), 7(1933), 80,133.

[4] E. F. Collingwood, Sur quelques théorèmes de M. Nevanlinna, CR Acad Sci, Paris, 179(1924), 955-957.

[5] C. T. Chuang, Uné generalisation d'une inégalité de Nevanlinna, Scientia Sinica,13(1964), 887-895.

[6] J. Dufresnoy, Sur les valeurs exceptionelles des fonctions meromorphes voisines d'une fonction meromorphe donnée, CR Acad. Sci, Paris, 208(1939), 255-257.

[7] H. Kneser, Zur Theorie der gebrochenen Funktionen mehrerer Veränderlichen, Jber. Deutsch. Math. Verein, 48(1938), 1-28.

[8] S. Mori, Remarks on holomorphic mappings, Contempory Math., 25(1983), 101-114.

[9] R. Nevanlinna, Zur Theorie der meromorphen Funktionen, Acta Math., 16(1925), 1-99.

[10] R. Nevanlinna, Le Théorème de Picard-Borel et la Théorie des Fonctions Méromorphes, Gauthiers-Villars, Paris, (1929), reprint Chelsea-Publ. Co., New York, (1974), pp 171.

[11] C. F. Osgood, Sometimes effective Thue-Siegel-Roth-Schmidt-Nevanlinna bounds, or better, to appear in Journal of Number Theory.

[12] C. F. Osgood, A fully general Nevanlinna N-small function theorem and a sometimes effective Thue-Siegel-Roth-Schmidt theorem for solutions to linear differential equations, Contemp. Math., 25(1983), 129-130.

[13] B. Shiffman, New defect relations for meromorphic functions on $\mathbb{C}^m$, Bull. Amer. Math. Soc. (New Series), 7(1982), 599-601.

[14] B. Shiffman, A general second main theorem for meromorphic functions on $\mathbb{C}$, Amer. J. Math., 106(1984), 509-531.

[15] W. Stoll, Ganze Funktionen endlicher Ordnung mit gegebenen Nullstellenflächen, Math. Zeitschrift, 57(1953), 211-237.

[16] W. Stoll, Die beiden Hauptsätze der Wertverteilungstheorie bei Funktionen mehrerer komplexen Veränderlichen, I Acta Math., 90(1953), 1-115; II Acta Math., 92(1954), 55-169.

[17] W. Stoll, Value distribution theory for meromorphic maps, Aspect of Math. E, 7(1985), pp 347 Springer-Verlag.

[18] A. Vitter, The lemma of the logarithmic derivative in several complex variables, Duke Math. J., 44(1977), 89-104.

[19] H. Weyl and J. Weyl, Meromorphic curves, Ann. of Math., 39(1938), 516-538.

[20] H. Weyl and J. Weyl, Meromorphic functions and analytic curves, Annals of Mathematics Studies, 12(1943), pp 269 Princeton University Press.

[21] H. Wu, The equidistribution theory of holomorphic curves, Annals of Mathematics Studies, 64(1970), pp 219 Princeton University Press.

[22] L. Yang, Deficient functions of meromorphic functions, Scientia Sinica, 24(1981), 1179-1189.

Vol. 1090: Differential Geometry of Submanifolds. Proceedings, 1984. Edited by K. Kenmotsu. VI, 132 pages. 1984.

Vol. 1091: Multifunctions and Integrands. Proceedings, 1983. Edited by G. Salinetti. V, 234 pages. 1984.

Vol. 1092: Complete Intersections. Seminar, 1983. Edited by S. Greco and R. Strano. VII, 299 pages. 1984.

Vol. 1093: A. Prestel, Lectures on Formally Real Fields. XI, 125 pages. 1984.

Vol. 1094: Analyse Complexe. Proceedings, 1983. Edité par E. Amar, R. Gay et Nguyen Thanh Van. IX, 184 pages. 1984.

Vol. 1095: Stochastic Analysis and Applications. Proceedings, 1983. Edited by A. Truman and D. Williams. V, 199 pages. 1984.

Vol. 1096: Théorie du Potentiel. Proceedings, 1983. Edité par G. Mokobodzki et D. Pinchon. IX, 601 pages. 1984.

Vol. 1097: R.M. Dudley, H. Kunita, F. Ledrappier, École d'Éte de Probabilités de Saint-Flour XII – 1982. Edité par P.L. Hennequin. X, 396 pages. 1984.

Vol. 1098: Groups – Korea 1983. Proceedings. Edited by A.C. Kim and B.H. Neumann. VII, 183 pages. 1984.

Vol. 1099: C.M. Ringel, Tame Algebras and Integral Quadratic Forms. XIII, 376 pages. 1984.

Vol. 1100: V. Ivrii, Precise Spectral Asymptotics for Elliptic Operators Acting in Fiberings over Manifolds with Boundary. V, 237 pages. 1984.

Vol. 1101: V. Cossart, J. Giraud, U. Orbanz, Resolution of Surface Singularities. Seminar. VII, 132 pages. 1984.

Vol. 1102: A. Verona, Stratified Mappings – Structure and Triangulability. IX, 160 pages. 1984.

Vol. 1103: Models and Sets. Proceedings, Logic Colloquium, 1983, Part I. Edited by G.H. Müller and M.M. Richter. VIII, 484 pages. 1984.

Vol. 1104: Computation and Proof Theory. Proceedings, Logic Colloquium, 1983, Part II. Edited by M.M. Richter, E. Börger, W. Oberschelp, B. Schinzel and W. Thomas. VIII, 475 pages. 1984.

Vol. 1105: Rational Approximation and Interpolation. Proceedings, 1983. Edited by P.R. Graves-Morris, E.B. Saff and R.S. Varga. XII, 528 pages. 1984.

Vol. 1106: C.T. Chong, Techniques of Admissible Recursion Theory. IX, 214 pages. 1984.

Vol. 1107: Nonlinear Analysis and Optimization. Proceedings, 1982. Edited by C. Vinti. V, 224 pages. 1984.

Vol. 1108: Global Analysis – Studies and Applications I. Edited by Yu.G. Borisovich and Yu.E. Gliklikh. V, 301 pages. 1984.

Vol. 1109: Stochastic Aspects of Classical and Quantum Systems. Proceedings, 1983. Edited by S. Albeverio, P. Combe and M. Sirugue-Collin. IX, 227 pages. 1985.

Vol. 1110: R. Jajte, Strong Limit Theorems in Non-Commutative Probability. VI, 152 pages. 1985.

Vol. 1111: Arbeitstagung Bonn 1984. Proceedings. Edited by F. Hirzebruch, J. Schwermer and S. Suter. V, 481 pages. 1985.

Vol. 1112: Products of Conjugacy Classes in Groups. Edited by Z. Arad and M. Herzog. V, 244 pages. 1985.

Vol. 1113: P. Antosik, C. Swartz, Matrix Methods in Analysis. IV, 114 pages. 1985.

Vol. 1114: Zahlentheoretische Analysis. Seminar. Herausgegeben von E. Hlawka. V, 157 Seiten. 1985.

Vol. 1115: J. Moulin Ollagnier, Ergodic Theory and Statistical Mechanics. VI, 147 pages. 1985.

Vol. 1116: S. Stolz, Hochzusammenhängende Mannigfaltigkeiten und ihre Ränder. XXIII, 134 Seiten. 1985.

Vol. 1117: D.J. Aldous, J.A. Ibragimov, J. Jacod, Ecole d'Ét Probabilités de Saint-Flour XIII – 1983. Édité par P.L. Henne IX, 409 pages. 1985.

Vol. 1118: Grossissements de filtrations: exemples et applicat Seminaire, 1982/83. Edité par Th. Jeulin et M. Yor. V, 315 pages. 1

Vol. 1119: Recent Mathematical Methods in Dynamic Programm Proceedings, 1984. Edited by I. Capuzzo Dolcetta, W.H. Fle and T. Zolezzi. VI, 202 pages. 1985.

Vol. 1120: K. Jarosz, Perturbations of Banach Algebras. V, 118 pa 1985.

Vol. 1121: Singularities and Constructive Methods for Their Treatr Proceedings, 1983. Edited by P. Grisvard, W. Wendland and Whiteman. IX, 346 pages. 1985.

Vol. 1122: Number Theory. Proceedings, 1984. Edited by K. A VII, 217 pages. 1985.

Vol. 1123: Séminaire de Probabilités XIX 1983/84. Proceedings. par J. Azéma et M. Yor. IV, 504 pages. 1985.

Vol. 1124: Algebraic Geometry, Sitges (Barcelona) 1983. Proc ings. Edited by E. Casas-Alvero, G.E. Welters and S. Xa Descamps. XI, 416 pages. 1985.

Vol. 1125: Dynamical Systems and Bifurcations. Proceedings, 1 Edited by B.L.J. Braaksma, H.W. Broer and F. Takens. V, 129 pa 1985.

Vol. 1126: Algebraic and Geometric Topology. Proceedings, 1 Edited by A. Ranicki, N. Levitt and F. Quinn. V, 523 pages. 1985

Vol. 1127: Numerical Methods in Fluid Dynamics. Seminar. Edite F. Brezzi, VII, 333 pages. 1985.

Vol. 1128: J. Elschner, Singular Ordinary Differential Operators Pseudodifferential Equations. 200 pages. 1985.

Vol. 1129: Numerical Analysis, Lancaster 1984. Proceedings. E by P.R. Turner. XIV, 179 pages. 1985.

Vol. 1130: Methods in Mathematical Logic. Proceedings, 1983. E by C.A. Di Prisco. VII, 407 pages. 1985.

Vol. 1131: K. Sundaresan, S. Swaminathan, Geometry and Nonli Analysis in Banach Spaces. III, 116 pages. 1985.

Vol. 1132: Operator Algebras and their Connections with Topo and Ergodic Theory. Proceedings, 1983. Edited by H. Araki, Moore, Ş. Strătilă and C. Voiculescu. VI, 594 pages. 1985.

Vol. 1133: K.C. Kiwiel, Methods of Descent for Nondifferentiable mization. VI, 362 pages. 1985.

Vol. 1134: G.P. Galdi, S. Rionero, Weighted Energy Methods in F Dynamics and Elasticity. VII, 126 pages. 1985.

Vol. 1135: Number Theory, New York 1983–84. Seminar. Edite D.V. Chudnovsky, G.V. Chudnovsky, H. Cohn and M.B. Nathan V, 283 pages. 1985.

Vol. 1136: Quantum Probability and Applications II. Proceedi 1984. Edited by L. Accardi and W. von Waldenfels. VI, 534 pa 1985.

Vol. 1137: Xiao G., Surfaces fibrées en courbes de genre deux. IX pages. 1985.

Vol. 1138: A. Ocneanu, Actions of Discrete Amenable Groups or Neumann Algebras. V, 115 pages. 1985.

Vol. 1139: Differential Geometric Methods in Mathematical Phy Proceedings, 1983. Edited by H. D. Doebner and J. D. Hennig. VI, pages. 1985.

Vol. 1140: S. Donkin, Rational Representations of Algebraic Gro VII, 254 pages. 1985.

Vol. 1141: Recursion Theory Week. Proceedings, 1984. Edite H.-D. Ebbinghaus, G.H. Müller and G.E. Sacks. IX, 418 pages. 1

Vol. 1142: Orders and their Applications. Proceedings, 1984. E by I. Reiner and K. W. Roggenkamp. X, 306 pages. 1985.

Vol. 1143: A. Krieg, Modular Forms on Half-Spaces of Quaterni XIII, 203 pages. 1985.

Vol. 1144: Knot Theory and Manifolds. Proceedings, 1983. Edite D. Rolfsen. V, 163 pages. 1985.